Dissertations
in
American Economic History

This is a volume in the Arno Press collection

Dissertations
in
American Economic History

Advisory Editor
Stuart Bruchey

Research Associate
Eleanor Bruchey

See last pages of this volume for a complete list of titles.

RELATIVE PRODUCTIVITY, FACTOR-INTENSITY AND TECHNOLOGY IN THE MANUFACTURING SECTORS OF THE U. S. AND THE U. K. DURING THE NINETEENTH CENTURY

Ephraim Asher

ARNO PRESS

A New York Times Company

New York / 1977

Editorial Supervision: LUCILLE MAIORCA

First publication in book form, Arno Press, 1977

DISSERTATIONS IN AMERICAN ECONOMIC HISTORY
ISBN for complete set: 0-405-09900-2
See last pages of this volume for titles.

Manufactured in the United States of America

Library of Congress Cataloging in Publication Data

Asher, Ephraim.
Relative productivity, factor-intensity and technology in the manufacturing sectors of the U. S. and the U. K. during the nineteenth century.

(Dissertations in American economic history)
Originally presented as the author's thesis, University of Rochester, 1969.
Includes bibliographical references.
1. Industrial productivity--United States--History. 2. Industrial productivity--Great Britain--History. 3. Production functions (Economic theory) 4. United States--Manufactures--History. 5. Great Britain--Manufactures--History. I. Title. II. Series.
HC110.I52A74 1977 338.4'5'0973 76-39822
ISBN 0-405-09902-9

RELATIVE PRODUCTIVITY, FACTOR-INTENSITY AND TECHNOLOGY IN THE MANUFACTURING SECTORS OF THE U. S. AND THE U. K. DURING THE NINETEENTH CENTURY

BY

EPHRAIM ASHER

Submitted in Partial Fulfillment

of the

Requirements for the Degree

DOCTOR OF PHILOSOPHY

Supervised by Professor Stanley Engerman

Department of Economics

The University of Rochester

Rochester, New York

1969

ABSTRACT

The main purpose of this study is to examine empirically the relationship between relative factor productivity and relative capital-intensity in the American and British manufacturing sectors during the nineteenth century. This problem is first formulated in theoretical terms. Subsequently the theoretical implications are used as a basis for the empirical results.

The study includes two parts. The first part analyzes the problem under the assumption that technology was the same in both countries. The second part examines the same question under the assumption that each country has different technology.

Chapter 1 in the first part postulates the problem in the framework of a neoclassical two-sector two-factor model incorporating a parameter of resource utilization. Subsequently we have derived relationships among relative factor-productivity, relative factor prices, and relative capital-intensity in both countries.

Chapters 2-4 in this part apply the foregoing theory to the cases of the main manufacturing branches - cotton, wool, and iron and steel. The period of enquiry covers most of the second half of the nineteenth century. The statistical data indicates that for two manufacturing sectors - cotton and wool - one may accept the presumption that the Americans used a more capital-intensive technique during the period in question. This could justify the corollary that relative labor-productivity was higher in the American sectors during the same period. In the case of iron and steel, the statistical evidence does not show a uniform trend. Here the data generally indicate that only after (roughly) 1870 capital intensity was higher in the U.S. Surrogates for relative utilization of natural resources in the case of textiles industries show higher values for the U.S. throughout the second half of the nineteenth century. This would be consistent with the contention that both labor and capital were more productive in the U.S. during this period of time.

In chapter 6 of the second part we have drawn theoretical foundations needed for the examination of the problem under the assumption of different technologies. We have adopted here production functions with constant returns to scale and constant elasticity of substitution (CES). This model enables estimation of factor-augmenting technological change. We have related this concept to the Hicksian notion of biased factor-saving technological change. Thus we were able to estimate the impact of relative changes in capital intensity on the relative changes in factor-productivities in both countries. The application of this model showed that both of the textiles industries in the U.S. were experiencing labor-saving bias during the second half of the nineteenth century. The results for the U.K. cotton industry do not reject the contention that capital-saving bias was experienced there during the same period. In the case of British wool industry there was labor-saving, rather than capital-saving, bias. We have concluded that the cotton industry in the U.S. was having a higher rate of growth of capital-intensity which contributed to a higher rate of growth of relative labor productivity as compared with the British sector. In the case of wool this result is reversed; namely, that the British sector was experiencing a higher rate of growth of capital-intensity, which meant a lower rate of growth of relative labor-productivity in the American sector during the second half of the nineteenth century.

As by-products of this study, we have obtained estimates for the elasticities of substitution for each of the textiles industries in both countries. The resulting estimates did not confirm the specification of constant unitary elasticity of substitution (Cobb-Douglas form) for either of the sectors in question. However, in the American cases, these estimates are quite low and may justify the use of production functions with zero elasticity of substitution.

TABLE OF CONTENTS

Table of Contents (cont.)

LIST OF TABLES

List of Tables (cont.)

List of Tables (cont.)

List of Tables (cont.)

List of Tables (cont.)

LIST OF DIAGRAMS

LIST OF CHARTS

INTRODUCTION

Very few economic historians deny that labor productivity was, and still is, higher in American industry than in British industry. Several attempts to estimate total labor productivity (output per unit of labor) show a higher ratio in the American manufacturing sector. Marvin Frankel in his "British and American Manufacturing Productivity" estimates that American manufacturing productivity was equal to that of Britain by about 1830, and has been higher ever since.[1] Based on Census data, A.W. Flux estimates that during 1924-25, American manufacturing net output per worker exceeded that of Britain by about 2.3 to 3.3.[2] L. Rostas estimates that total labor productivity in American industry was more than twice as high as that in the United Kingdom in the pre-war period of 1935-39.[3] D. Paige and G. Bombach show that for 1950, average labor productivity in manufacturing was at least two and a half times as high in the U.S. as in Britain.[4]

---

1. Frankel, Marvin, "British and American Manufacturing Productivity", University of Illinois Bulletin, Urbana, 1957, p. 29. This is a highly tentative estimate subject to a wide margin of error, but it gives an indication of the historical nature of the problem.

2. Flux, W.W., "Industrial Productivity in Great Britain and the United States", *QJE*, November 1933, p. 17.

3. Rostas, L., *Comparative Productivity in British and American Industry*, Cambridge University Press, 1948, p. 27.

4. Paige, Deborah and Gottfried Bombach, *A Comparison of National Output and Productivity of the United Kingdom and the United States*, OEEC, Paris 1959, p. 55.

The question which immediately follows is what caused this "superior efficiency" of American industry. Several attempts were made to explain it, and we shall briefly discuss a few of them. One explanation was the relatively larger geographical size of the U.S. which resulted in a larger market and stimulated large scale production and higher factor productivity.[5] Another emphasized the special social structure in the U.S. which resulted in a higher demand for standardized goods which in return led to a more efficient and productive economic organization.[6] The leading theory, however, is the one which associated the higher American industrial efficiency with the structure of resource and factor availability. This theory, which was developed first by E. Rothbarth and further by H.J. Habakkuk, holds that a different factor endowment in the U.S. vis-a-vis the U.K. resulted in a higher capital-intensity and therefore, in a relatively higher labor productivity.[7] In brief, (see

---

5. For an early summary and criticism of these views, see E. Rothbarth "Causes of the Superior Efficiency of U.S.A. Industry as Compared with British Industry", EJ, September 1946.

6. See Harold F. Williamson, "Mass Production, Mass Consumption and American Industrial Development", First International Conference of Economic History, Stockholm, 1960.

7. See Rothbarth, *op.cit.*; Also see H.J. Habakkuk, *American and British Technology in the Nineteenth Century: The Search for Labor Saving Inventions*, Cambridge, 1962.

detailed explanation later), the essential point was that relatively higher availability of land and greater scarcity of labor in the U.S. led to labor-saving inventions and techniques. This process ultimately resulted in a higher capital-intensity and hence in a higher labor productivity in the U.S.

Apart from theoretical questions that the Rothbarth-Habakkuk argument may raise[8], there are two basic empirical questions that one may wish to examine:

a) Was American manufacturing experiencing labor-saving technological progress as compared with Britain?

b) Were processes of production in American manufacturing more capital-intensive than in British manufacturing?

These are the basic questions which will be examined in our study.

The setting up of a theoretical framework within which one may examine these questions will raise, inter-alia, the question of what to assume about the state of technologies in both countries.

---

8. Theoretically, there is a basic question whether relative labor-scarcity in one country necessarily implies labor-saving technological change there. In his two-factor model, Fellner proposes that one cannot conclude a priori whether relative labor-scarcity will result in labor-saving technological change. However, his first proposition implies that if relative labor-scarcity at "present" will induce people to expect a rise in real wage-rates relative to interest rates in the "future", firms under competitive conditions will prefer labor-saving technological progress. [See William Fellner, "Two Propositions in the Theory of Induced Innovation", EJ, June 1961, p. 306.]

This question, as we shall see later, is quite crucial to the argument. In our study we shall employ with question (a) the assumption that technologies were different in both countries. Here it is simply assumed that different biased-innovations will result in different technologies in both countries. Under this assumption, question (a), as we shall see, may become one of dynamics; namely, a question of the rate of growth of factor efficiency. Question (b), on the other hand, will be examined under the assumption of identical technologies in both countries. Under this assumption one can establish uniquely-defined relationships between factor-intensity in both countries and their relative factor-productivity. Question (b) under this assumption, however, becomes one of comparative statics; namely, a question of relative capital-intensity at a given point of time. Before we continue to say more on this subject, we would like to mention at this stage, some of the problems which were involved in previous attempts to deal with this question.

Without a great deal of difficulty, one can trace in the basic explanation of Rothbarth a "skeleton" of an economic model and subsequently build an explicit one which will enable us to examine question (b) under the assumption of identical technologies. "In any country," says Rothbarth, "where land is readily available in large quantities, labor is likely to be expensive. For the income of the industrial worker must be sufficiently high to present an alternative to his cultivating the land for his own profit.

Thus the high productivity of labor in American industry at the beginning of this century can be explained by the fact that industry had to install labor-saving equipment and to economize in the use of labor until its productivity was sufficiently far higher than it was in agriculture to enable relatively attractive wages to be paid in industry."[9]

Reading this explanation carefully, one may grant that the following assumptions are not necessarily in contrast with the tacit economic model that Rothbarth had in mind:

a. in the American economy there are two sectors — industry and agriculture;

b. the production function in each sector involves two factors — land and labor in agriculture, and labor and capital in industry;

c. factors are internationally immobile;

d. trade is not absolutely free;

e. the economic system is competitive.

Under these assumptions, the logic of Rothbarth's argument becomes quite simple. Given that land was abundant in the U.S., a relatively high land-labor ratio was prevalent in agriculture. From this a relatively high marginal productivity of labor follows. Under the assumption of a competitive system, this will ultimately equal the real-wage-rate of labor and therefore will provide an opportunity cost for labor in its alternative use in industry. It follows that real remuneration for labor in industry was at least the same as in

---

9. Rothbarth, op.cit., p. 385.

agriculture. This was accomplished by installing relatively higher capital-intensive techniques.[10]

Peter Temin, in an attempt to examine the problem for the period of the 1850's, suggested a theoretical model based on the same grounds.[11] In his model he adopts the basic assumptions (a) - (e). Adding to them the basic assumptions of the neoclassical two-sector model, his fundamental question becomes the following: Given that technology was identical in both countries, was it possible that the American manufacturing sector was using more capital-intensive processes? He argues that under the conditions of his model, higher capital-intensity in the U.S. would have implied a lower interest rate. During the 1850's, however, rates of interest were higher in the U.S.; therefore, labor-intensive rather than capital-intensive techniques must have been used in the American manufacturing sector.[12]

---

10. Alternatively, one can use the following simple model in order to demonstrate the nature of the problem.

Assume two competitive systems with two sectors in each as follows:

$$(1)\quad X_A^i = F^i(L,T)\ ,$$

$$(2)\quad X_M^i = G^i(L,K)\ , \qquad i=1,2.$$

(1) and (2) are production functions of agriculture and industry respectively. Let the X^1's stand for outputs in the U.S., and the X^2's for outputs in the U.K. The factors of production are assumed to be homogeneous but internationally immobile. Assume also that artificial and/or real impediments to free-trade do exist. F and G are assumed to be homogeneous of the first degree, single-valued with continuous positive first- and negative second-order partial derivatives. It follows that

---

$$X_A^i = TF^i(\frac{L}{T},1) ,$$
$$= Tf^i(\frac{L}{T}) , \qquad i = 1,2.$$

so that

$$MP_{LA}^i \equiv \frac{\partial F^i}{\partial L} = \frac{df^i(\frac{L}{T})}{d(\frac{L}{T})} ;$$

where MP_{LA} is the marginal productivity of labor in agriculture.

Now translating into the terminology of this model the contention of Rothbarth and Habakkuk about relative labor-scarcity in the U.S., we have

$$(\frac{L}{T})^1 < (\frac{L}{T})^2$$

It follows then that

$$\frac{df^1(\frac{L}{T})}{d(\frac{L}{T})} > \frac{df^2(\frac{L}{T})}{d(\frac{L}{T})} ;$$

i.e., $MP_{LA}^1 > MP_{LA}^2$.

Under the assumption of competitive equilibrium, wages(in terms of final manufacturing output, i.e., using manufacturing output as numeraire) should have been higher in the American manufacturing sector because in agriculture wages were higher and hence they provided an opportunity cost to industrial laborers. But higher wages in the American manufacturing sector could have prevailed only if

$$MP_{LM}^1 > MP_{LM}^2 ,$$

i.e., only if

$$\frac{dg^1(\frac{L}{K})}{d(\frac{L}{K})} > \frac{dg^2(\frac{L}{K})}{d(\frac{L}{K})}$$

where $g^i(\frac{L}{K}) = \frac{1}{K} G^i(L,K) , \qquad i=1,2.$

But the latter inequality implies that (L/K) in the U.S. was lower than in the U.K., i.e., that American manufacturing was using processes with higher capital-intensity.

11. Temin, P., "Labor Scarcity and the Problem of American Industrial Efficiency in the 1850's", JEH, September 1966. See also a note by Ian M. Drummond in this Journal, September 1967, and a reply by Temin in the same Journal, March 1968.

---

12. In the context of the model given in Note (10), one can state Temin's argument as follows:

Assume that X_M is used for both consumption and capital formation. Abstracting from depreciation and risk, the interest rate, r, will be given by

$$r^i = \frac{\partial X_M^i}{\partial K} = \frac{dq^i(\frac{K}{L})}{d(\frac{K}{L})}, \qquad i=1,2;$$

where $$q^i(\frac{K}{L}) = \frac{1}{L} G^i(L,K), \qquad i=1,2.$$

Now the fact that during the 1850's the interest rate in the U.S. was higher means simply that

$$\frac{dq^1(\frac{K}{L})}{d(\frac{K}{L})} > \frac{dq^2(\frac{K}{L})}{d(\frac{K}{L})}$$

which implies that (K/L) was lower in the U.S.

According to Temin, it follows that under identical technologies, relative labor productivity (in terms of marginal physical product), was lower in the U.S. than in the U.K. This is indeed in contrast with the "orthodox" view of the superiority of American industrial efficiency. Also, as a corollary, his conclusion with respect to capital-intensity is in contrast with that of Rothbarth and Habakkuk.

Temin's results are undoubtedly the consequence of specification errors. Robert Fogel, in his "The Specification Problem in Economic History",[13] points out that, in the first place, the manufacturing production function in Temin's model should have included the factor "land" (or natural resources). In the second place, technologies may not have been the same because of structural differences within the industrial sectors of both countries. Thus, according to Fogel, Temin does not eliminate the possibility that the U.S. was both labor- and capital-scarce. He adds that even if American and British technologies were identical, the fact that interest rates were higher in the U.S. does not necessarily imply that "real-wages" were lower. In Fogel's model this can be the case only if British agricultural prices were below those in the U.S. But since the opposite was observed, namely, that agricultural prices in the U.S. were lower, he concludes that it was

---

13. Fogel, Robert W., "The Specification Problem in Economic History", JEH, September 1967.

possible that both American interest rates and "real-wages" exceeded those which prevailed in the U.K. during the ante-bellum era.

Indeed, one can attribute differences in processes of production in manufacturing to differences in technologies of the two countries.[14] However, it is still desirable to examine the problem of comparative factor-intensity (and hence the question of relative factor productivity) in the manufacturing sectors under the assumption of identical technologies. This can still give an indication of the problem of factor substitution and relative factor-costs. This question will be examined in the first part of this study. A modified version of Temin's model incorporating a "third factor" into the manufacturing production function will be used. Subsequently we shall bring statistical data, covering the period of the second half of the nineteenth century to show the implications for the structures of capital intensity in three manufacturing sub-sectors. We may anticipate, at this stage, that the principal conclusions reached via this route are in accord with the traditional view.

In the second part of this study we shall employ the second assumption that technologies were different in both

---

14. In a joint article E. Ames and N. Rosenberg analyze the establishment of the Enfield Arsenal in the context of Temin's approach. They point out that differences in processes between the two countries may have been the result of differences in technology and/or non-homogeneity of inputs and outputs, and not only the result of differences in relative factor-prices. (See E. Ames and N. Rosenberg, "The Enfield Arsenal in Theory and History", *EJ*, December 1968).

countries. As we mentioned before, the problem under this assumption becomes one of trying to find evidence of factor saving bias in technological change in both countries. Following Habakkuk's contention of relative labor-scarcity in the U.S. (and assuming that Fellner's first proposition holds), one might expect labor-saving bias in its manufacturing sector. A corollary of this can be that as a result of relative labor-abundancy in the U.K., one might expect capital-saving bias in its manufacturing sector. We shall specify an aggregate production function which will allow us to measure a biased technological change. We shall, then, attempt to measure this bias in both countries. We may say that the results achieved through this approach were generally supporting (or at least not fully contradicting) the conclusions implied by Rothbarth and Habakkuk. The statistical data, and hence the resulting evidence, in this part also cover mainly the period of the second half of the nineteenth century.

CHAPTER I

The Model

The model which will be described below is basically the same as the one given in Note (10). It is stated here in order to lay down the necessary theoretical foundations for our forthcoming empirical study. This theoretical basis should be kept in mind because it is essential to the evaluation of the empirical results.

Consider again two competitive systems with two sectors in each.

Let

(1) $X^i_M = \alpha^i_M F^i(L,K)$,

(2) $X^i_A = G^i(L,T)$, $i=1,2;$

be the production functions of the two sectors in both systems.

In this notation X^1_M and X^1_A stand for outputs of manufacturing and agricultural sectors respectively in the U.S.; X^2_M and X^2_A are the same for the U.K.: α^1_M is a scalar-index measuring the degree of actual utilization of natural resources for the manufacturing sector in the U.S.; α^2_M is the same for the U.K.; L, K and T are homogeneous labor, capital, and land respectively.

Assume that F and G are homogeneous of the first degree, single-valued with continuous positive first- and negative second-order partial derivatives.

Assume that imperfect factor mobility and/or imperfect free trade do exist between the two countries. This assumption enables us to explain factor-price differentials between the two countries. Thus, we exclude the case of factor-price equalization which may evolve from our theoretical setup.

X_M is assumed to be used for either consumption or capital. At this stage we will disregard depreciation.

We assume that the same sector in both countries has the same technology. Diagrammatically we are saying that each sector has the same map of isoquants. Since we exclude externalities, we can simply define "identical technologies" in terms of identical elasticities of substitutions. In other words, we can assume at this point that the same sector has the same elasticity of substitution for any given capital-labor-ratio.[15]

Equation (1) states that the manufacturing output is a function of two variable factors — labor and capital. The utilization of natural resources is treated as a parameter. This formulation, as we shall see below, will enable us to derive a clear relationship between relative capital-intensity and relative factor productivity. If natural resources had been included as a third factor in the manufacturing production function, we would not have been able to derive such a relationship (see further

---

15. Different technologies are defined, then, in terms of different elasticities of substitution for a given capital-labor-ratio. Diagrammatically, this implies different maps of isoquants.

elaboration on the significance of this formulation in Note (21).

It follows from (1) that

$$X_M^i = \alpha_M^i \, LF^i(1, \frac{K}{L}) \; ,$$

$$(3) \quad \therefore \; X_M^i = L\alpha_M^i f^i(k) \; , \qquad i=1,2;$$

and $k \equiv \frac{K}{L}$

Also

$$X_M^i = \alpha_M^i \, KF^i(\frac{L}{K}, 1) \; ,$$

$$= K\alpha_M^i \, q^i(\frac{1}{k}) \; ,$$

$$(4) \quad \therefore \; X_M^i = K\alpha_M^i \, g^i(k) \; , \qquad i=1,2.$$

In equilibrium, the real wage rate, w, and the interest rate, r, will be equated to marginal physical productivities of labor and capital respectively:[16]

$$w^i = \frac{\partial X_M^i}{\partial L} = \alpha_M^i \frac{dg^i}{dk} \; ,$$

$$r^i = \frac{\partial X_M^i}{\partial K} = \alpha_M^i \frac{df^i}{dk} \; , \qquad i=1,2.$$

---

16. In equilibrium, real-wage-rates (say in terms of the manufacturing good) will be equal in both sectors. Thus, for the purpose of our study, we can disregard equation (2) and proceed with equation (1). Indeed, one can state the problem in terms solely of the production function in the manufacturing sector, as is done by Temin.

Given the assumptions of this model, one can easily verify the following relationships at equilibrium:

$$(5)\qquad \frac{w^1}{r^1} \gtrless \frac{w^2}{r^2} \iff k^1 \gtrless k^2 .$$

The proof of (5) is quite simple. In each country

$$(6)\begin{cases} w = \alpha g'(k) , \\ g''(k) > 0 , \qquad\qquad \text{and let} \\ \left.\begin{array}{l} g'(0) = 0 \\ g'(\infty) = \infty \end{array}\right\} \end{cases}$$

Also

$$(7)\begin{cases} r = \alpha f'(k) , \\ f''(k) < 0 , \qquad\qquad \text{and let} \\ \left.\begin{array}{l} f'(0) = \infty \\ f'(\infty) = 0 \end{array}\right\} \end{cases}$$

Given (6) and (7) one can easily proceed to prove (5). We shall prove only the first of these relationships because the proofs of the other two are straightforward:

$$(5a)\qquad \frac{w^1}{r^1} > \frac{w^2}{r^2} \iff k^1 > k^2 .$$

Proof:

if $\qquad \dfrac{w^1}{r^1} > \dfrac{w^2}{r^2}$;

then by substitution we get

$$\frac{\alpha_M^1 \dfrac{dg^1(k)}{dk}}{\alpha_M^1 \dfrac{df^1(k)}{dk}} > \frac{\alpha_M^2 \dfrac{dg^2(k)}{dk}}{\alpha_M^2 \dfrac{df^2(k)}{dk}}$$

It follows that

$$\frac{\frac{dg^1(k)}{dk}}{\frac{dg^2(k)}{dk}} > \frac{\frac{df^1(k)}{dk}}{\frac{df^2(k)}{dk}}$$

This inequality is possible only if $k^1 > k^2$.

Then we have

$$\frac{\frac{dg^1(k)}{dk}}{\frac{dg^2(k)}{dk}} > 1 \ ,$$

and

$$\frac{\frac{df^1(k)}{dk}}{\frac{df^2(k)}{dk}} < 1 \ .$$

These follow from the relationships in (6) and (7). Of course, these inequalities will be reversed for $k^1 < k^2$ and become equalities for $k^1 = k^2$. Thus, we proved the necessary condition for the relationships of (5a). The proof of the second half of (5a) follows immediately again from (6) and (7), so we shall not consider it here.

The relationships in (5) are the basic rules that we shall employ for our empirical study. Under these rules the problem of capital-intensity in the manufacturing sector of each system becomes a question of relative factor-prices.[17] These rules

---

17. We assume here equilibrium factor prices. In equilibrium, however, factor prices are simultaneously determined by many variables, including the prevailing demand conditions for the final manufacturing output. Thus in this sense demand conditions

---

enter the picture in determining the relative capital intensity. In other words, different relative demand for X_M may determine, inter-alia, different factor price ratios, and hence different capital intensity. Relative capital intensity remains, however, a function of relative factor prices.

enable us to examine the question of capital-intensity by simply examining the structure of factor-prices in both countries. In the subsequent chapters of this part we shall employ these rules to draw the main conclusions. In particular we shall use the necessary conditions for the relationships in (5) as theoretical guides to each of the subsectors in question.[18] The conclusions which will be drawn from each subsector will be used in return as the basis for possible generalization about the manufacturing sector as a whole.

The economic logic behind the necessary conditions in (5) is that under identical technologies and at equilibrium, one country uses relatively more of one factor when it costs relatively less. Thus it is sufficient to know the relative factor-prices in order to know the relative factor-intensity in each country under these conditions. (Diagrammatically, this simply means that a different

---

18. Since, in our study, we examine three sectors in each country rather than aggregate for one manufacturing sector, we could have stated the production function under (1) in the following disaggregation manner:

$$(1a)\qquad X^{i}_{M_j} = \alpha^{i}_{M_j}\, F(L_j, K_j)\ , \qquad i=1,2;\quad j=a,b,c;$$

where X_{M_j} is the manufacturing output of the jth subsector (a,b,c). Under this approach, each subsector may use different type of labor and different type of capital, but in both systems in the jth output, homogeneity of factors is assumed. In order to get the same theoretical implications, one must, however, assume that the jth output goes either to consumption or to formation of K_j which makes it quite a strong a-sumption. One may also notice that the w's and r's under (1a) will become marginal physical productivity of labor and "own-interest-rate" for the jth subsector, respectively.

slope of isocost line determines a different process of production on the same isoquant curve.[19]

The model proposed by Temin does not allow that both factors are more efficient in the same country. If, for example, interest rates were higher in the U.S. (and hence marginal physical productivity of capital), it must follow that marginal physical productivity of labor was lower in the same period. Alternatively, if capital-labor-ratio in both countries is different, then marginal productivities of both factors cannot exceed in one country those of the other. (It is possible, however, to have identical marginal productivities of each factor in both countries when relative capital-intensity is the same in both.)

---

19. Thus, the necessary conditions of the relationships under (5a) imply the following diagram:

Diagram A: Determination of Capital-Intensity Under Perfect Competition

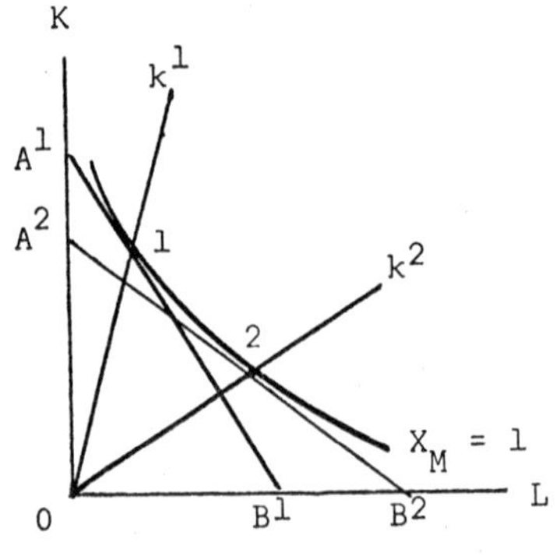

The fact that $\frac{w^1}{r^1} > \frac{w^2}{r^2}$ means simply that the slope of $\overline{A^1B^1}$ is greater than the slope of $\overline{A^2B^2}$ (the isocost lines of country 1 and country 2 respectively). At equilibrium $k^1 > k^2$.

As in Fogel's model,[20] our model allows that one country can have superior efficiency in both factors. This is possible now because of the new element of α_M in the production function.[21]

---

20. Fogel, *op.cit.*, p. 301.

21. The main purpose of this study is to examine the relationship between relative factor productivity and relative factor intensity in the American and British manufacturing sectors. The main theoretical problem involved here is to construct a model which will enable us to make empirical examination of this question under the assumption of identical technologies. For this purpose one can use the neoclassical two-sector, two-factor model used so frequently in economic theory. To use this model, however, would have necessarily involved the specification error already committed by Temin, namely, that both countries face an identical "factor-price" frontier. In order to avoid such an error, one could have used a three-factor manufacturing production function such as the one proposed by Fogel (see Fogel, *op.cit.*, p. 301) as follows:

$$(1^1) \qquad X_M^i = k^i\ (L,K,T)\ , \qquad i=1,2$$

and where T = land (or natural resources).

The introduction of the third factor, T, into the production function would have allowed both countries to have different "factor-price" frontiers. However, it would have eliminated the assumption that marginal productivities are functions of capital-labor-ratio with homogeneity of degree zero. In order to retain this assumption, and at the same time not to commit Temin's error, we have introduced the concept of α_M. One can look upon α_M as an alternative way of incorporating Fogel's third factor of "land" into the neoclassical two-factor production function (thus different levels of α_M will correspond to different "factor'price" frontiers.) Now, however, we can still retain the assumption of zero homogeneity between marginal productivities (of L and K) and the capital-labor-ratio (K/L). Furthermore, since Part II of the study deals with the relationships between relative factor productivities and capital intensity under the assumption of biased technological change, one may indeed say that our model incorporates the case of neutral technological progress in both countries. Because in our formulation, a difference between the levels of α_M's will indicate a difference between the levels of combined-factor-efficiency, as a result of different (neutral) technological progress (leaving relative marginal factor productivities unaltered.)

The scalar α_M is a theoretical concept behind one's "common-sense" that relative richness in natural resources may affect relative factor productivity. The basic idea is that each country differs in its potential wealth in natural resources and hence in its actual utilization for the manufacturing sector.[22] A relatively higher potential-richness and hence actual utilization of natural resources will increase factor productivities, assuming all other things are given. The α_M's, then, represent here the impact on the productivities of capital and labor as a result of differences in potential and hence actual utilization of natural resources in the U.S. and the U.K. We note that the absolute level of α_M has no impact on the shape of the production function. From this point of view, the notion of the α_M's is very similar to the notion of neutral technological change in the production function.

In this system, then, the following relationships hold:

$$(8) \quad \left.\begin{array}{l} w^1 \gtreqless w^2 \\ r^1 \gtreqless r^2 \end{array}\right\} \Longrightarrow \alpha^1 \gtreqless \alpha^2 ,$$

for any k^1 and k^2.

---

22. Of course, actual utilization of natural resources in the social organization is a function of several variables including inter-alia, the general level of economic activity, the political structure, the sociological and cultural environment and relations with other social organizations. Given, however, all the other factors, one should expect higher level of resource utilization the higher the potential richness or availability in natural resources.

The proofs here again are quite simple and we shall consider the first case given by (8a) as follows:

$$\text{(8a)} \qquad \left.\begin{array}{l} w^1 > w^2 \\ r^1 > r^2 \end{array}\right\}, \Longrightarrow \alpha^1 > \alpha^2$$

for any k^1 and k^2.

Proof:

By rearrangement and substitution we get

$$\left.\begin{array}{l} w^1 > w^2 \\ r^1 > r^2 \end{array}\right\}, \Longrightarrow \left\{\begin{array}{l} \text{I.} \dfrac{\frac{dg^1(k)}{dk}}{\frac{dg^2(k)}{dk}} > \dfrac{\alpha^2}{\alpha^1} \\[2ex] \text{II.} \dfrac{\frac{df^1(k)}{dk}}{\frac{df^2(k)}{dk}} > \dfrac{\alpha^2}{\alpha^1} \end{array}\right.$$

We distinguish among three cases as follows:

a. if $k^1 = k^2$, then the simultaneous inequalities in I and II will hold only if $\alpha^1 > \alpha^2$ (because the left term of these inequalities equals unity);

b. if $k^1 > k^2$, then the left term of I is greater than unity but the left term of II is less than unity, so the simultaneous inequalities in I and II will hold only if $\alpha^1 > \alpha^2$;

c. if $k^1 < k^2$, then the left term of I is less than unity but that of II is greater than unity, so the only way for these inequalities to hold is when $\alpha^1 > \alpha^2$.

Thus (8a) implies that $\alpha^1 > \alpha^2$ no matter what the capital-labor-ratio is in each country.

It is also true that a higher α_M in one country does not necessarily imply that both factors are more efficient there. A higher α_M in one country is a necessary condition for both factors to be more productive there, but it is not a sufficient condition.[23]

We shall make use of the relationships under (8) in our empirical study by constructing special surrogates for the concept of α_M.

---

23. Different α_M in each country means, simply, that the factor price frontier in each country is different. In Temin's model there is only one factor price frontier common to both countries. Here, although both have the same technology, the country with the greater α_M will have the higher frontier. If, for example, $\alpha_M^1 > \alpha_M^2$ then diagrammatically the factor price frontier of country 1 (F^1F^1) is higher than that of country 2 (F^2F^2). This situation is shown by Diagram B, as follows:

Diagram B: Factor Price Frontiers and the Size of α_M.

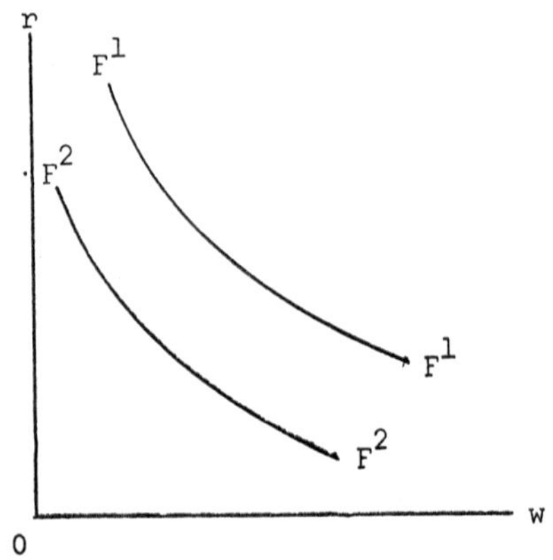

CHAPTER 2

The Structure of Factor Prices and its Implications for the Actual Capital-Intensity in the Major Manufacturing Branches

The main question that we are trying to answer in this chapter is: can the actual structure of factor prices in the major branches of industry tell us something about the "implied" actual capital-intensity in both countries? In other words, if actual relative factor-prices show repeatedly that in one branch one of the factors was relatively cheaper, then the implied conclusion would be that this factor was used relatively more in that branch during that period. This, of course, follows from our model (relationships (5)).

One of the major dilemmas confronting those who apply theoretical models is how to use the available statistics of our complicated world in order to follow the theory as closely as possible. The model requires that factor-prices would be given in terms of marginal physical products (MPP's). The MPP of capital in the model, however, is identical with the interest rate. Thus we have taken as surrogate for "r" interest rate series. It would have been ideal to find "own-interest-rates" for each of the branches, but because of lack of data, we have used the same series of interest-rates for each branch.[24]

---

24. Tentative estimates of the rate of return on capital in textiles, based on U.S. census data, show higher rates as compared with long term interest rates. Lack of comparable data for the U.K. prevented the use of such series as surrogates for "r". (These estimates, however, define the rate of return on capital as a ratio between the total share of capital and total value of capital used in a given year.)

Another question is what kind of interest rates one may use as a proxy. Indeed, the results to a larger extent will depend on the type of interest rates used. We shall draw our main conclusions on the basis of long-term interest rates. But we shall also compare these results with series of short term interest rates, which will be given for this matter in the Appendix (Table A1).

The case of surrogates for the "w's" was much easier to tackle. Here we have calculated two alternative series as surrogates for the MPP of labor in each branch. First, by deflating money-wage-rates by the price of a "given commodity".[25] Money-wage-rates are treated as surrogates for value of the MPP of labor in equilibrium. Since, however, in our model the concept of "w" stands also for "real-wage", one may argue that its surrogate should be the usual definition of real-wages, namely, money-wages deflated by the cost of living index. Thus, following this definition, we have calculated alternative series of surrogates for the MPP of labor. We shall employ both of these definitions. The main conclusions will be drawn from the first type of these series, but we shall also mention the types of conclusions that one may draw by using series of the "usual" definition of real-wages. The latter, however, will be given in the Appendix (Tables A3-A16).

---

25. This is a surrogate for the manufacturing output "M". The commodities which are being used in this study as surrogates are major final-outputs in each branch. However, they are not necessarily produced directly by each and every occupation taken as surrogate for labor.

Thus, if we want to know whether the American sector was more capital-intensive, we have to look at statistical evidence showing that the interest-"real-wage"-ratio was lower in the U.S. Equivalently, the same conclusion will follow if we observe that the American-British interest-rate-ratio was lower than the American-British "real-wage"-ratio.[26] Now because all branches in one country are represented by the same series of interest-rates, we shall calculate ratios of interest rates and then compare them with "real-wage"-ratios of each branch in the two countries.

A. Interest Rates in the U.S. and the U.K. During the Nineteenth Century

Table 1 shows that during most of the nineteenth century, long-term interest rates in the U.S. did exceed those of the U.K. Except for the period between 1882 and 1891, the ratio between the American and the British rates was greater than unity. This ratio was considerably high during the period of the Civil War. Using this data, a follower of Temin's approach may hastily suggest that not only during the 1850's one should have observed lower capital-intensity in the U.S., but also throughout the entire nineteenth century except for the short period indicated above.[27] We shall see, however, that this was not quite so. But let us look first at the picture as a whole. This is given in Table 1 and Chart 1 as follows:

---

26. Simply, because $(\frac{w^1}{r^1}) > (\frac{w^2}{r^2})$ implies $(\frac{w^1}{w^2}) > (\frac{r^1}{r^2})$.

27. It should be noted that different changes in the price levels in both countries result in differentials in the "real" interest rate over time. This problem was disregarded for the sake of simplicity.

TABLE 1: Long Term Interest Rates in the U. S. and the U. K. During the Nineteenth Century (1800-1900)[28]

%

Year	Interest Rates (1) U.S.[29]	(2) U.K.[30]	Ratio (1)/(2)	Year	Interest Rates (1) U.S.[29]	(2) U.K.[30]	Ratio (1)/(2)
1800	6.94	4.71	1.47	1820	5.16	4.42	1.47
1801	6.44	4.92	1.30	1821	4.57	4.07	1.12
1802	6.02	4.23	1.42	1822	4.65	3.79	1.22
1803	6.16	4.99	1.23	1823	4.72	3.80	1.24
1804	6.29	5.30	1.18	1824	4.25	3.30	1.28
1805	6.38	5.04	1.26	1825	4.32	3.54	1.22
1806	6.14	4.87	1.26	1826	4.50	3.79	1.18
1807	6.08	4.92	1.23	1827	4.37	3.61	1.21
1808	5.96	4.55	1.30	1828	4.48	3.54	1.26
1809	5.85	4.49	1.30	1829	4.50	3.34	1.34
1810	5.82	4.47	1.30	1830	4.37	3.49	1.25
1811	5.95	4.67	1.27	1831	4.41	3.76	1.17
1812	6.12	5.08	1.20	1832	4.45	3.58	1.24
1813	6.30	4.92	1.28	1833	4.87	3.42	1.42
1814	7.64	4.92	1.55	1834	4.87	3.32	1.46
1815	7.30	4.48	1.63	1835	4.83	3.29	1.46
1816	7.25	5.02	1.44	1836	4.96	3.35	1.48
1817	5.86	4.10	1.43	1837	4.95	3.30	1.50
1818	5.78	3.87	1.49	1838	5.01	3.23	1.55
1819	5.90	4.17	1.41	1839	5.21	3.28	1.58

TABLE 1: (continued)

%

Year	Interest Rates (1) U.S.	(2) U.K.	Ratio (1)/(2)	Year	Interest Rates (1) U.S.	(2) U.K.	Ratio (1)/(2)
1840	5.07	3.35	1.51	1862	6.25	3.23	1.93
1841	4.99	3.38	1.47	1863	6.00	3.24	1.85
1842	6.07	3.27	1.85	1864	5.10	3.33	1.53
1843	5.03	3.17	1.58	1865	5.19	3.35	1.55
1844	4.85	3.03	1.60	1866	5.17	3.41	1.51
1845	5.16	3.12	1.65	1867	4.97	3.23	1.53
1846	5.50	3.13	1.75	1868	4.62	3.20	1.44
1847	5.77	3.44	1.67	1869	4.07	3.23	1.26
1848	5.71	3.51	1.62	1870	4.24	3.24	1.30
1849	5.16	3.24	1.59	1871	4.18	3.23	1.29
1850	4.58	3.11	1.47	1872	3.70	3.24	1.14
1851	4.47	3.09	1.44	1873	3.51	3.24	1.08
1852	4.39	3.02	1.45	1874	3.42	3.24	1.05
1853	4.82	3.07	1.30	1875	3.30	3.20	1.03
1854	4.14	3.27	1.26	1876	3.66	3.16	1.15
1855	4.18	3.31	1.26	1877	3.81	3.15	1.20
1856	4.11	3.22	1.27	1878	3.97	3.15	1.26
1857	4.30	3.27	1.31	1879	3.96	3.08	1.28
1858	4.32	3.10	1.39	1880	3.63	3.05	1.19
1859	4.72	3.15	1.49	1881	3.13	3.00	1.04
1860	5.57	3.19	1.74	1882	2.91	2.99	0.97
1861	6.45	3.28	1.96	1883	2.88	2.97	0.96

TABLE 1: (concluded)

Year	Interest Rates (1) U.S.	Interest Rates (2) U.K.	Ratio (1)/(2)	Year	Interest Rates (1) U.S.	Interest Rates (2) U.K.	Ratio (1)/(2)
1884	2.76	2.97	0.93	1893	2.96	2.61	1.13
1885	2.68	3.02	0.88	1894	2.72	2.52	1.08
1886	2.43	2.98	0.81	1895	2.82	2.39	1.18
1887	3.32	2.95	1.12	1896	3.06	2.28	1.34
1888	2.27	2.97	0.76	1897	2.57	2.25	1.14
1889	2.13	2.81	0.75	1898	2.50	2.28	1.10
1890	2.37	2.67	0.88	1899	2.22	2.36	0.94
1891	2.58	2.70	0.95	1900	2.58	2.54	1.01
1892	2.73	2.65	1.03				

---

28. The source is S. Homer, A History of Interest Rates, Rutgers University Press, New Jersey, 1963.

29. These are calculated for *actual* market yields on long-term U.S. Federal Government Bonds. For the decade of 1833-42, we took New England Municipal Bond yields as an approximation for Federal Bonds because of absence of the former type of data. See Homer, *ibid.*, pp. 268-88.

30. These are actual yields on long-term British Consols (3%). See Homer, *ibid.*, pp. 195-97.

Chart 1: American and British Interest Rates

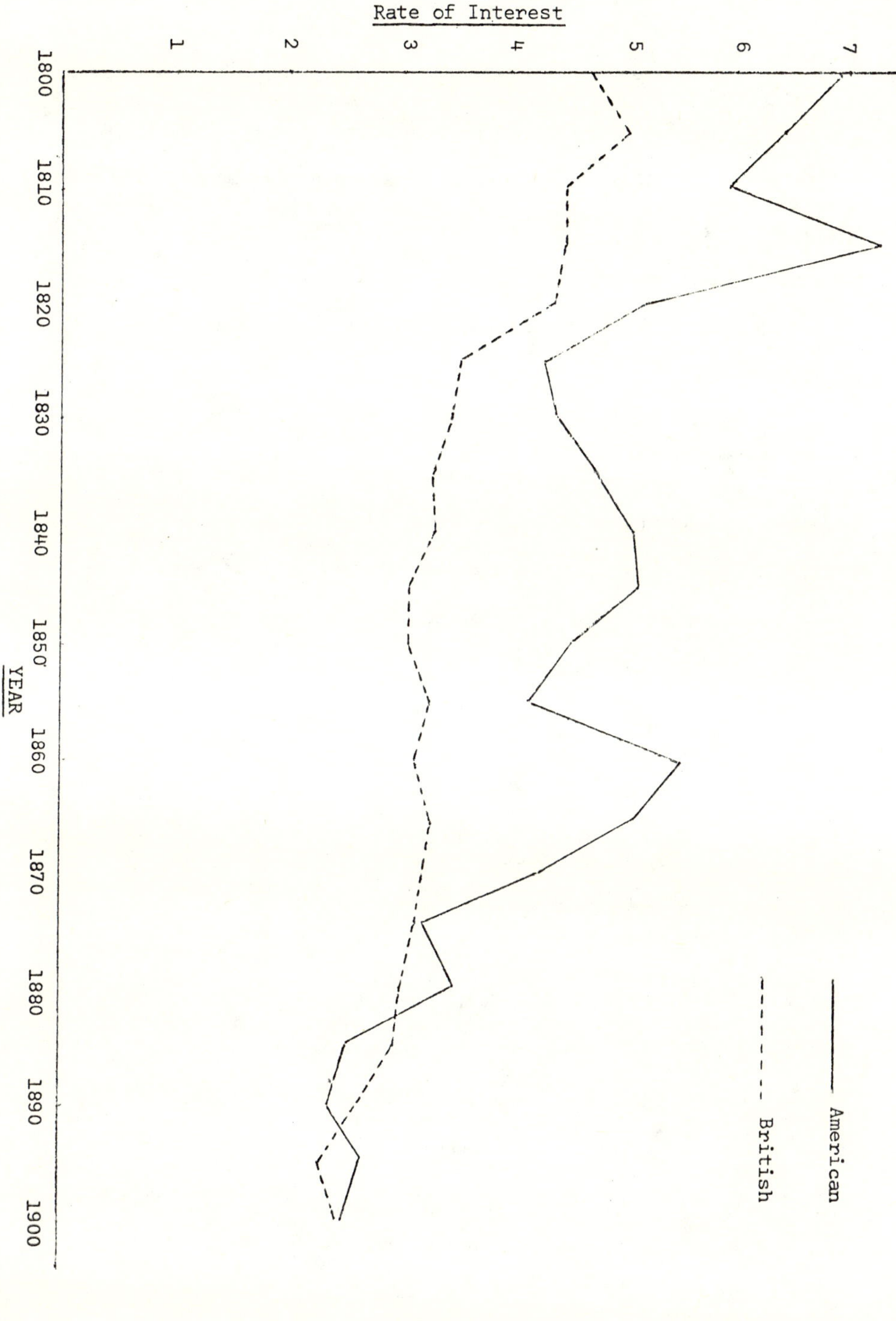

The next step will be to compare the results under column (3) in Table 1 with American-British "real-wage"-ratios. If the latter were below the former, we may expect that under the same technology, manufacturers in the U.S. should have been encouraged to use capital-saving processes. We shall see that this was not necessarily the case.

To demonstrate this, series of "real-wages" (or surrogates for the MPP of labor) from three major sectors in both countries were calculated and are presented here. These branches are: cotton, wool, and iron and steel. Overall averages as well as representative occupation averages will be presented.

In the following sections we shall present "real-wages" in terms of a given manufacturing product. As indicated above, however, we shall also refer to the second type of "real-wages", namely, money-wages deflated by the cost of living index.

In both cases the units of labor are measured by weeks of work, so that wages are given in terms of average weekly earnings.

We tried to cover as much as possible of the nineteenth century. The availability of data (indeed, the lack of its availability) restricted the study to the period between 1840 and 1900. However, the period for most occupations is shorter. It was not an easy task to find data of identical nature for both countries, but we tried to come as close as possible to it. This concerned mainly the problem of finding wage series in both countries of workers performing the same job.[31] Nevertheless, we were able to find four such occupations for each branch. The variety of these jobs covers skilled as well as unskilled labor. For two sectors — cotton and wool — there was enough data to construct overall averages as well.[32] The wool averages, however, will be given in the Appendix only.

B. "Real-Wages" in the Cotton Industry in the U.S. and the U.K. During the Nineteenth Century

The wage series in this section covers the period from 1848 to 1893. They include overall averages (1848-1893); unskilled labor which is represented by "Piecers" (1850-1886); more skilled labor, "Grinders" (1850-1886); and most skilled workers which are represented by two types of overseers (1849-1886).

These series are given in Tables 2-6 as follows:

---

31. The purpose of selecting occupations of identical nature is to make comparisons of "real-wages" in the context of the model more meaningful. We note, however, that non-homogeneity of labor as well as products does exist between the two countries.

32. The weights of different occupations in these averages are not identical for the two countries. See explanations and sources for these averages in the Appendix.

TABLE 2: Average Weekly Earnings in the Cotton Industry in the U.S. and the U.K. During the Nineteenth Century (1848-1893): Overall Averages[33]

Year	Nominal Earnings U.S. $(1)	Nominal Earnings U.K. Pence(2)	Av. Price of Cloth per Yd. U.S. ¢(3)	Av. Price of Cloth per Yd. U.K. Pence(4)	Cloth per Worker per Week U.S. Yds.(5)	Cloth per Worker per Week U.K. Yds.(6)	Ratio (5)/(6)
1848	3.76	125.74	4.35	2.29	86.44	54.90	1.58
1849	3.68	130.92	4.58	2.50	80.35	52.37	1.53
1850	3.81	125.43	5.19	2.88	73.41	43.55	1.69
1853	3.75	135.42	6.15	2.78	60.98	48.71	1.25
1854	3.75	124.88	5.81	2.60	64.54	48.03	1.34
1859	3.64	141.55	5.67	3.04	64.20	46.56	1.38
1860	3.71	140.22	5.44	3.13	68.20	44.80	1.52
1865	5.20	129.75	20.24	5.35	25.69	24.25	1.06
1870	6.01	155.61	7.14	3.41	84.17	45.63	1.84
1871	6.07	158.20	7.41	3.04	81.92	52.04	1.57
1873	6.47	163.72	6.69	3.17	96.71	51.65	1.87
1874	5.96	171.01	5.57	3.06	107.00	55.89	1.91
1876	5.53	167.04	4.10	2.67	134.88	62.56	2.16
1877	5.13	193.38	4.38	2.74	117.12	70.58	1.66
1880	4.89	187.75	4.51	2.52	108.43	74.50	1.46
1882	5.47	196.04	3.76	2.50	145.48	78.42	1.86
1883	5.53	192.86	3.60	2.32	153.61	83.13	1.85
1884	5.38	196.39	3.36	2.30	160.12	85.39	1.88
1886	5.31	183.68	3.31	2.17	160.42	84.65	1.90
1893	6.30	196.73	3.30	2.35	190.91	83.71	2.28

---

33. The sources for the series are as follows:

Column (1): Layer, Robert G. and Arthur H. Cole, Earning of Cotton Mill Operatives 1825-1914, Committee on Research in Economic History. Harvard University Press, 1955. The averages given by Layer and Cole are average daily earnings of male workers. To obtain these averages, they have divided total earnings of all workers in the selected mills by total number of days worked by all workers. This method gave, essentially, equal weights to different occupations. To find the weekly earnings, these numbers were multiplied by six.

Column (2): Bowley, A.L., Wages in the United Kingdom in the Nineteenth Century, Cambridge University Press. To obtain overall averages, we have used the wage series for spinners, piecers and weavers with 1886's weights, p. 118.

Column (3): U.S. Bureau of the Census, Statistical Abstract of the U.S., 1895. Government Printing Office, Washington, D.C., 1896, p. 374. These are annual averages for printing-cloth (or gray cloth) in New York.

Column (4): Blaug, M., "The Productivity of Capital in the Lancashire Cotton Industry During the Nineteenth Century". EHR, Vol. XIII (1960-61), p. 376. To find the average price per yard, Blaug's prices were divided by the average number of yards per pound (approximately 4.6).

TABLE 3: Average Weekly Earnings in the Cotton Industry in the U.S. and the U.K. During the Nineteenth Century (1850-1886): Piecers[34]

Year	Nominal Earnings U.S. $(1)	U.K. Pence(2)	Av. Price of Cloth per Yd. U.S. ¢(3)	U.K. Pence(4)	Cloth per Worker per Week U.S. Yds.(5)	U.K. Yds.(6)	Ratio (5)/(6)
1850	2.01	78	5.19	2.88	38.73	27.08	1.43
1859	2.64	72	5.67	3.04	46.56	23.68	1.96
1860	2.64	78	5.44	3.13	48.53	24.92	1.94
1870	4.07	96	7.14	3.41	57.00	28.15	2.02
1874	4.38	96	5.57	3.06	78.63	31.37	2.50
1877	3.96	102	4.38	2.74	90.41	37.22	2.42
1880	3.18	102	4.51	2.52	70.51	40.48	1.74
1882	3.65	132	3.76	2.50	97.07	52.80	1.83
1884	3.75	96	3.36	2.30	111.61	41.74	2.67
1886	3.60	117	3.31	2.17	108.76	53.91	2.01

---

34. The sources are as follows:

Column (1): Aldrich, Mr., *Reports of Committees of the Senate of the U.S. for the 2nd Session of the 52nd Congress, March, 1893.* Government Printing Office, Washington, D.C., 1893. Part 3, pp. 842-43. These are averages for January and July from Establishment No. 41 New York.
Column (2): Bowley, *Wages in the U.K.*, *op.cit.*, p. 118. These are averages for Manchester and District.
Columns (3)-(4): See Note (33).

TABLE 4: Average Weekly Earnings in the Cotton Industry in the U.S. and the U.K. During the Nineteenth Century (1850-1886): Grinders[35]

Year	Nominal Earnings U.S. $(1)	Nominal Earnings U.K. Pence(2)	Av. Price of Cloth per Yd. U.S. ¢(3)	Av. Price of Cloth per Yd. U.K. Pence(4)	Cloth per Worker per Week U.S. Yds.(5)	Cloth per Worker per Week U.K. Yds.(6)	Ratio (5)/(6)
1850	4.89	168	5.19	2.88	94.22	58.33	1.62
1853	5.64	184	6.15	2.78	91.71	66.18	1.38
1854	5.64	168	5.81	2.60	97.10	55.26	1.75
1859	5.88	180	5.67	3.04	103.70	59.21	1.75
1860	6.15	183	5.44	3.13	113.05	58.47	1.93
1865	8.61	192	20.24	5.35	42.49	35.89	1.18
1870	10.92	218	7.14	3.41	152.90	64.00	2.39
1874	9.24	282	5.57	3.06	165.90	92.15	1.80
1877	7.65	265	4.38	2.74	174.64	96.71	1.80
1880	8.34	270	4.51	2.52	184.92	107.14	1.72
1882	7.68	252	3.76	2.50	204.25	100.80	2.02
1883	8.31	253	3.60	2.32	230.83	109.05	2.11
1884	8.43	233	3.36	2.30	250.89	101.31	2.47
1886	9.00	257	3.31	2.17	271.90	118.43	2.38

---

35. The sources are as follows:

Column (1): Aldrich Reports, *op.cit.*, Part 2, pp. 562-66. Establishment No. 38, Massachusetts.
Column (2): Bowley, *Wages in the U.K.*, *op.cit.*, p. 118.
Columns (3)-(4): See Note (33).

TABLE 5: Average Weekly Earnings in the Cotton Industry in the U.S. and the U.K. During the Nineteenth Century (1849-1886): Overseers, Spinning Department[36]

Year	Nominal Earnings U.S. \$(1)	Nominal Earnings U.K. Pence(2)	Av. Price of Cloth per Yd. U.S. ¢(3)	Av. Price of Cloth per Yd. U.K. Pence(4)	Cloth per Worker per Week U.S. Yds.(5)	Cloth per Worker per Week U.K. Yds.(6)	Ratio (5)/(6)
1849	12.00	264	4.58	2.50	262.01	105.60	2.48
1850	11.25	264	5.19	2.88	216.76	91.67	2.36
1859	12.00	312	5.67	3.04	211.64	102.63	2.06
1860	12.00	312	5.44	3.13	220.59	99.68	2.21
1870	10.50	360	7.14	3.41	147.06	105.57	1.40
1873	15.72	360	6.69	3.17	234.98	113.56	2.06
1877	18.00	480	4.38	2.74	410.95	175.18	2.34
1880	25.50	480	4.51	2.52	565.41	190.47	2.96
1882	22.50	510	3.76	2.50	598.40	204.00	2.93
1883	24.00	600	3.60	2.32	666.67	258.62	2.57
1886	24.60	510	3.31	2.17	743.20	235.02	3.16

---

36. The sources are as follows:

Column (1): Aldrich Reports, *op.cit.*, Part 3, pp. 657-59, Establishment No. 39, Massachusetts.
Column (2): Bowley, *Wages in the U.K.*, *op.cit.*, p. 118.
Columns (3)-(4): See Note (33).

TABLE 6: Average Weekly Earnings in the Cotton Industry in the U.S. and the U.K. During the Nineteenth Century (1849-1886): Overseers, Carding Department[37]

Year	Nominal Earnings U.S. $(1)	Nominal Earnings U.K. Pence(2)	Av. Price of Cloth per Yd. U.S. ¢(3)	Av. Price of Cloth per Yd. U.K. Pence(4)	Cloth per Worker per Week U.S. Yds.(5)	Cloth per Worker per Week U.K. Yds.(6)	Ratio (5)/(6)
1849	12.00	336	4.58	2.50	262.01	134.40	1.94
1850	12.00	324	5.19	2.88	231.21	112.50	2.05
1859	15.00	336	5.67	3.04	263.62	110.53	2.38
1860	15.00	336	5.44	3.13	275.74	107.35	2.57
1870	18.00	384	7.14	3.41	252.10	112.61	2.24
1873	18.00	384	6.69	3.17	269.06	121.14	2.22
1874	18.00	450	5.57	3.06	323.15	147.06	2.19
1877	18.00	474	4.38	2.74	410.96	172.99	2.37
1880	24.00	462	4.51	2.52	532.15	183.34	2.90
1882	30.00	528	3.76	2.50	797.87	211.20	3.77
1883	30.00	492	3.60	2.32	833.34	212.07	3.92
1886	30.00	468	3.31	2.17	906.34	215.67	4.20

---

37. The sources are as follows:

Column (1): Aldrich Reports, _op.cit._, Part 3, p. 769, Establishment No. 40, Massachusetts.
Column (2): Bowley, _Wages in the U.K._, _op.cit._, p. 118.
Columns (3)-(4): See Note (33).

Chart 2: American-British MPP-of-Labor Ratios in the Cotton Industry

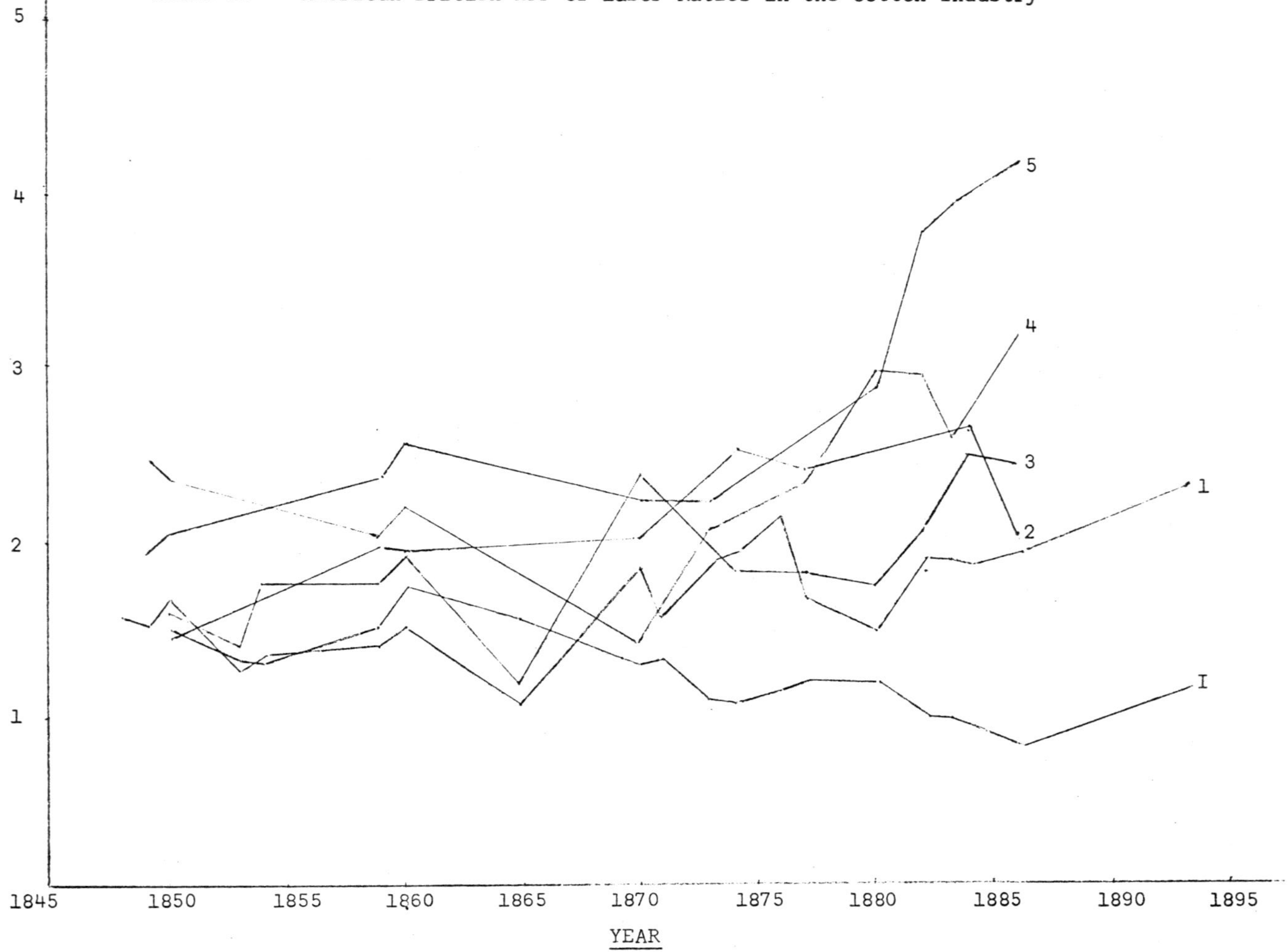

LEGEND:

1. Overall Averages
2. Piecers
3. Grinders
4. Overseers, Spinning Department
5. Overseers, Carding Department
I. American-British Interest Rate Ratios

In all cases "real-wages" are given in terms of gray cloth, namely, nominal weekly earnings were deflated by the price of cloth per yard to yield "yards-per-worker-per-week" (columns (5) and (6)). These can be viewed, then as the MPP of labor per week.[38]

By looking at these tables, one can easily see that throughout the period in question and without exception, these surrogates for the MPP of labor were higher in the U.S.

The ratios in the last column in each of these tables give us the second and most important result for our argument. These are surrogates for American-British MPP-of-labor (or American-British "real-wage"-ratios). If we compare them with American-British interest-rate-ratios, we can see that in most of the period the former ratios were higher (see Chart 2). To be more specific, we can see that — except for the years 1853, 1859, 1860, and 1865 in Table 2, the year 1850 in Table 3, and the year 1865 in Table 4 — the rest of the period shows that for all occupations, labor was relatively more "productive" in the American sector. This result is particularly strong in the case of more skillful workers — overseers in carding and spinning.

C. "Real-Wages" in the Woolen Industry in the U.S. and the U.K. During the Nineteenth Century

In this case, as in the previous one, representative occupations are taken. However, because of lack of price-data,

---

38. We assume, of course, that labor is paid its value of marginal product. By assuming so, however, one cannot avoid disregarding the problem of non-competitive elements which exist in the real world.

we were unable to construct overall averages here. For the same reason, the period is shorter and covers the years 1871-1891. Again, unskilled as well as skilled labor series are presented here. These are given in the following order: "Washers" (1871-1886), Dyers (1871-1886), "Sorters" (1871-1891), and Overseers in the Weaving Department (1871-1886).

Nominal weekly earnings were deflated by the price of flannels per yard. Thus, surrogates for the MPP of labor are given in terms of flannels per worker per week.

These series are given in Tables 7-10 as follows:

TABLE 7: Average Weekly Earnings in the Woolen Industry in the U.S. and the U.K. During the Nineteenth Century (1871-1886): Washers[39]

Year	Nominal Earnings		Av. Price of Flannels per Yd.		Flannels per Worker per Week		Ratio
	U.S. $(1)	U.K. Pence(2)	U.S. $(3)	U.K. Pence(4)	U.S. Yds.(5)	U.K. Yds.(6)	(5)/(6)
1871	7.50	198	0.475	17.55	15.79	11.28	1.40
1872	7.50	240	0.515	17.65	14.56	13.60	1.07
1874	6.96	204	0.440	19.76	15.82	10.32	1.53
1876	6.78	264	0.360	17.52	18.83	15.07	1.25
1877	6.00	216	0.336	17.58	17.86	12.29	1.45
1883	6.75	270	0.361	14.82	18.70	18.22	1.03
1886	6.99	245	0.330	12.49	21.18	19.62	1.08

---

39. The sources are as follows:

Column (1): Aldrich Reports, *op.cit.*, Part 4, pp. 1554-55, Establishment No. 88, Rhode Island.

Column (2): Bowley, A.L., "The Statistics of Wages in the U.K. During the Last Hundred Years",Part IX, *JRSS*, 1902. pp. 104-05.

Column (3): Aldrich Reports, *op.cit.*, Part 2, pp. 139-40.

Column (4): Aldrich Reports, *ibid.*, Part 1, p. 281. These, however, are wholesale export prices which were usually lower than wholesale domestic prices.

TABLE 8: Average Weekly Earnings in the Woolen Industry in the U.S. and the U.K. During the Nineteenth Century (1871-1886): Dyers[40]

Year	Nominal Earnings U.S. $(1)	Nominal Earnings U.K. Pence(2)	Av. Price of Flannels per Yd. U.S. $(3)	Av. Price of Flannels per Yd. U.K. Pence(4)	Flannels per Worker per Week U.S. Yds.(5)	Flannels per Worker per Week U.K. Yds.(6)	Ratio (5)/(6)
1871	9.00	198	0.475	17.55	18.95	11.28	1.68
1872	9.00	204	0.515	17.65	17.48	11.56	1.51
1874	9.00	288	0.440	19.76	20.46	14.58	1.40
1876	9.00	264	0.360	17.52	25.00	15.07	1.66
1877	6.72	264	0.336	17.58	20.00	15.02	1.33
1886	9.00	233	0.330	12.49	27.27	18.66	1.46

---

40. The sources are as follows:

Column (1): Aldrich Reports, *ibid.*, Part 4, pp. 1523-24, Establishment No. 88, Massachusetts.
Column (2): Bowley, "The Statistics of Wages in the U.K. During the Last Hundred Years", *JRSS*, 1902, *op. cit.*, pp. 104-05.
Columns (3)-(4): See Note (39).

TABLE 9: Average Weekly Earnings in the Woolen Industry in the U.S. and the U.K. During the Nineteenth Century (1871-1891): Sorters[41]

Year	Nominal Earnings U.S. $(1)	Nominal Earnings U.K. Pence(2)	Av. Price of Flannels per Yd. U.S. $(3)	Av. Price of Flannels per Yd. U.K. Pence(4)	Flannels per Worker per Week U.S. Yds.(5)	Flannels per Worker per Week U.K. Yds.(6)	Ratio (5)/(6)
1871	13.02	300	0.475	17.55	27.41	17.09	1.60
1872	13.11	336	0.515	17.65	25.46	19.04	1.34
1874	12.93	274	0.440	19.76	29.39	13.87	2.12
1876	11.85	324	0.360	17.52	32.92	18.49	1.78
1877	12.15	384	0.336	17.58	36.16	21.84	1.66
1883	15.21	370	0.361	14.82	42.13	24.97	1.69
1886	16.50	340	0.330	12.49	50.00	27.22	1.84
1891	12.24	360	0.301	12.42	40.66	28.99	1.40

---

41. The sources are as follows:

Column (1): Aldrich Reports, *op.cit.*, Part 4, pp. 1488-90, Establishment No. 86, Connecticut.
Column (2): Bowley, "The Statistics of Wages in the U.K. During the Last Hundred Years", *JRSS*, 1902, *op.cit.*, pp. 104-05.
Columns (3)-(4): See Note (39).

TABLE 10: Average Weekly Earnings in the Woolen Industry in the U.S. and the U.K. During the Nineteenth Century (1871-1886): Overseers, Weaving Department[42]

Year	Nominal Earnings U.S. $(1)	U.K. Pence(2)	Av. Price of Flannels per Yd. U.S. $(3)	U.K. Pence(4)	Flannels per Worker per Week U.S. Yds.(5)	U.K. Yds.(6)	Ratio (5)/(6)
1871	18.00	348	0.475	17.55	37.90	19.83	1.91
1874	21.00	408	0.440	19.76	47.73	20.65	2.31
1876	24.00	396	0.360	17.52	66.67	22.60	2.95
1877	18.00	432	0.336	17.58	53.57	24.57	2.18
1883	21.00	408	0.361	14.82	58.17	27.53	2.11
1886	18.00	372	0.330	12.49	54.55	29.78	1.83

---

42. The sources are as follows:

Column (1): Aldrich Reports, op.cit., Part 4, p. 1510, Establishment No. 87, Massachusetts.
Column (2): Bowley, "The Statistics of Wages in the U.K. During the Last Hundred Years", JRSS, 1902, op.cit., pp. 104-05.
Columns (3)-(4): See Note (39).

Chart 3: American-British MPP-of-Labor Ratios in the Woolen Industry

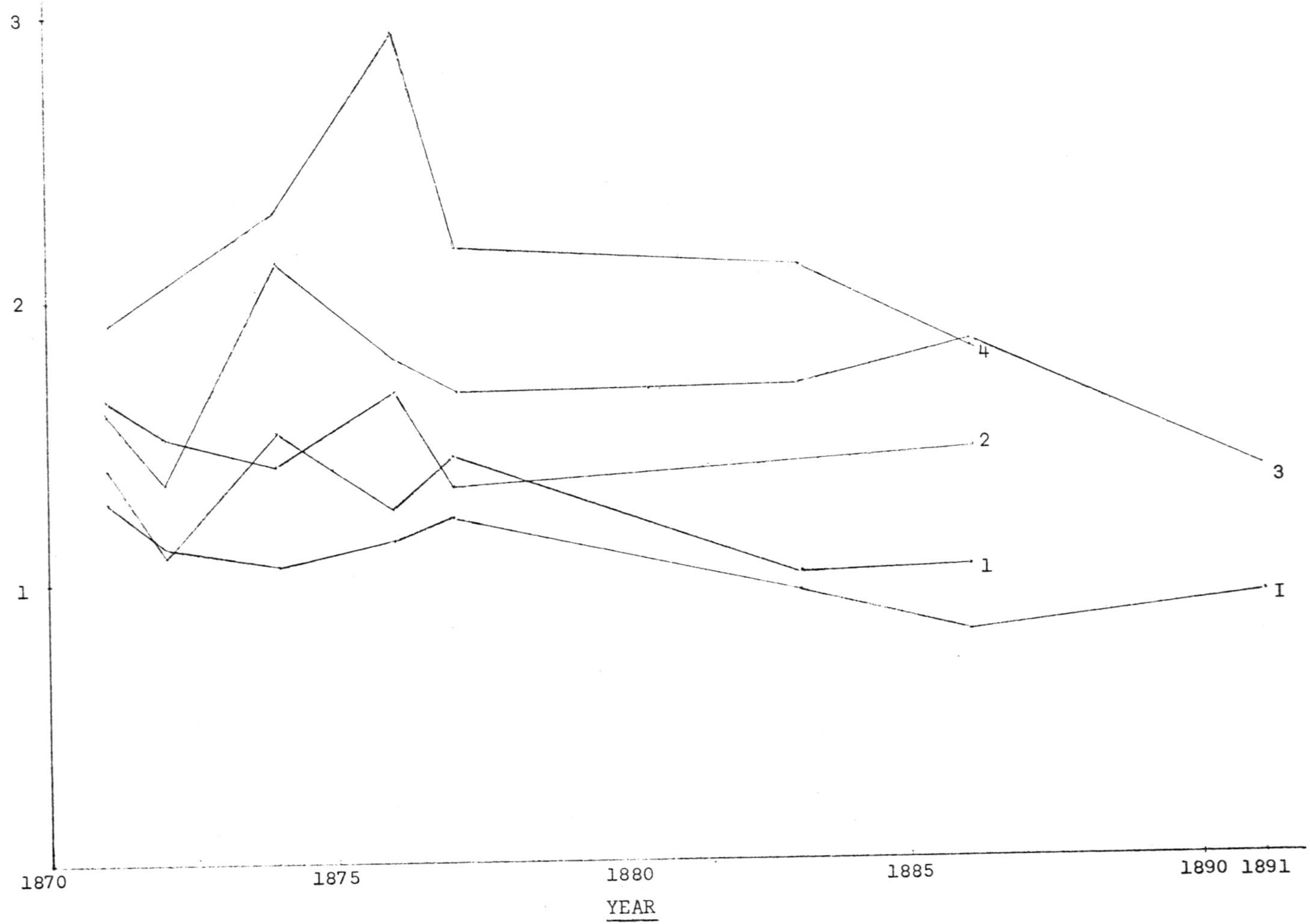

LEGEND:

1. Washers
2. Dyers
3. Sorters
4. Overseers, Weaving Department

I. American-British Interest Rate Ratios

We can see again that surrogates for the MPP of labor were higher in the U.S. throughout the period in question and for all occupations. Also we can see that with the exception of 1872 in Table 7, American-British MPP-of-labor-ratios were higher than the American-British interest-rate-ratios (see Chart 3). Again this phenomenon was stronger in the case of more skilled labor and especially in the case of overseers.

D. Real-Wages in the Iron and Steel Industry in the U.S. and the U.K. During the Nineteenth Century

The period which is covered here is between 1845 and 1886. We were able to find data for two price-deflators, namely, bar-iron and pig-iron. The results which were obtained via these price deflators are quite similar so we shall bring in this section only those obtained by deflating with bar-iron prices, while the results with pig-iron prices can be found in the Appendix (Tables A17-A20).

The occupations covered are "machinists" (1845-1886), "millwrights" (1859-1886), "boilermakers" (1859-1886), and pattern-makers (1849-1886).[43] These series are given in Tables 11-14 as follows:

---

43. These occupations do not participate in the production of pig-iron. However, they are classified under the branch of iron and steel. They are selected for our purpose to preserve homogeneity of labor in this sector.

TABLE 11: Average Weekly Earnings in the Iron and Steel Industry in the U.S. And the U.K. During the Nineteenth Century (1845-1886):

Machinists[44]

Year	Nominal Earnings U.S. $(1)	Nominal Earnings U.K. s/ d(2)	Av. Price of Bar Iron per Ton U.S. $(3)	Av. Price of Bar Iron per Ton U.K. L/ s/d(4)	Iron per Worker per Week (tons) U.S. (5)	Iron per Worker per Week (tons) U.K. (6)	Ratio (5)/(6)
1845	8.97	29/ 7	93.75	6/10/0	0.096	0.228	0.42
1846	9.51	29/ 7	91.66	9/10/0	0.104	0.156	0.67
1849	11.40	23/10	67.50	6/ 5/0	0.169	0.190	0.89
1871	19.20	34/ 0	78.54	7/12/6	0.244	0.223	1.09
1874	19.50	34/ 0	67.95	10/ 5/0	0.287	0.166	1.73
1884	16.68	32/ 0	44.05	5/ 2/6	0.379	0.312	1.21
1886	16.68	32/10	43.12	4/12/6	0.391	0.355	1.10

---

44. The sources are as follows:

Column (1): Aldrich Reports, *op.cit.*, Part 4, pp. 1277-82, Establishment No. 70, New York.

Column (2): Bowley, *Wages in the U.K.*, *op.cit.*, p. 122.

Column (3): U.S. Bureau of the Census, *Statistical Abstract of the U.S. 1890*, Government Printing Office, Washington, D.C., 1891, p. 343.

Column (4): Mitchell, B.R. and P. Deane, *Abstract of British Historical Statistics*. Cambridge University Press, London, 1962, p. 493.

TABLE 12: Average Weekly Earnings in the Iron and Steel Industry in the U.S. and the U.K. During the Nineteenth Century (1859-1886):

Millwrights[45]

Year	Nominal Earnings U.S. $(1)	U.K. s/d(2)	Av. Price of Bar Iron per Ton U.S. $(3)	U.K. L/ s/d(4)	Iron per Worker per Week (tons) U.S. (5)	U.K. (6)	Ratio (5)/(6)
1859	12.00	32/0	60.00	6/15/0	0.200	0.237	0.84
1871	21.00	30/0	78.54	7/12/6	0.267	0.196	1.36
1877	16.50	32/0	45.55	6/15/0	0.362	0.237	1.52
1880	15.00	31/0	60.38	6/15/0	0.248	0.229	1.08
1884	16.80	34/0	44.05	5/ 2/6	0.381	0.331	1.15
1886	15.00	33/6	43.12	4/12/6	0.348	0.362	0.96

---

45. The sources are as follows:

Column (1): Aldrich Reports, *op.cit.*, Part 4, pp. 1357-58, Establishment No. 74, Pennsylvania.
Column (2): Bowley, *Wages in the U.K.*, *op.cit.*, p. 118.
Columns (3)-(4): See Note (44).

TABLE 13: Average Weekly Earnings in the Iron and Steel Industry in the U.S. and the U.K. During the Nineteenth Century (1859-1886):

Boilermakers[46]

Year	Nominal Earnings U.S. $(1)	U.K. s/d(2)	Av. Price of Bar Iron per Ton U.S. $(3)	U.K. L/ s/d(4)	Iron per Worker Per Week (tons) U.S. (5)	U.K. (6)	Ratio (5)/(6)
1859	12.48	32/0	60.00	6/15/0	0.208	0.237	0.87
1860	11.64	30/0	58.75	6/10/0	0.198	0.230	0.86
1864	13.38	33/6	146.46	8/ 5/0	0.091	0.203	0.45
1870	14.22	30/0	78.96	7/ 7/6	0.180	0.203	0.88
1874	18.99	35/8	67.95	10/ 5/0	0.279	0.174	1.60
1877	14.04	34/0	45.55	6/15/0	0.308	0.251	1.23
1883	17.49	32/0	50.30	5/15/0	0.347	0.278	1.24
1884	17.49	33/8	44.05	5/ 2/6	0.397	0.328	1.21
1886	17.49	32/2	43.12	4/12/6	0.405	0.348	1.16

---

46. The sources are as follows:

Column (1): Aldrich Reports, op.cit., Part 3, pp. 1060-62, Establishment No. 57, Maryland.
Column (2): Bowley, Wages in the U.K., op.cit., p. 122.
Columns (3)-(4): See Note (44).

TABLE 14: Average Weekly Earnings in the Iron and Steel Industry in the U.S. and the U.K. During the Nineteenth Century (1849-1886):

Patternmakers[47]

Year	Nominal Earnings U.S. $(1)	Nominal Earnings U.K. s/d(2)	Av. Price of Bar Iron per Ton U.S. $(3)	Av. Price of Bar Iron per Ton U.K. L/ s/d(4)	Iron per Worker per Week (tons) U.S. (5)	Iron per Worker per Week (tons) U.K. (6)	Ratio (5)/(6)
1849	9.63	30/0	67.50	6/ 5/0	0.142	0.240	0.59
1851	10.38	30/6	54.66	5/10/0	0.190	0.277	0.69
1859	11.10	32/0	60.00	6/15/0	0.185	0.237	0.78
1864	15.93	31/6	146.46	8/ 5/0	0.109	0.191	0.57
1871	20.46	37/0	78.54	7/12/6	0.260	0.242	1.07
1877	15.78	31/0	45.55	6/15/0	0.346	0.229	1.51
1880	15.90	30/0	60.38	6/15/0	0.263	0.222	1.18
1883	20.34	35/0	50.30	5/15/0	0.404	0.304	1.33
1884	21.00	37/0	44.05	5/ 2/6	0.477	0.361	1.32
1886	19.62	34/8	43.12	4/12/6	0.455	0.375	1.21

---

47. The sources are as follows:

Column (1): Aldrich Reports, op.cit., Part 4, pp. 1309-12, Establishment No. 71, New York.
Column (2): Bowley, Wages in the U.K., op.cit., p. 122.
Columns (3)-(4): See Note (44).

Chart 4: American-British MPP-of-Labor Ratios in the Metallic Industry

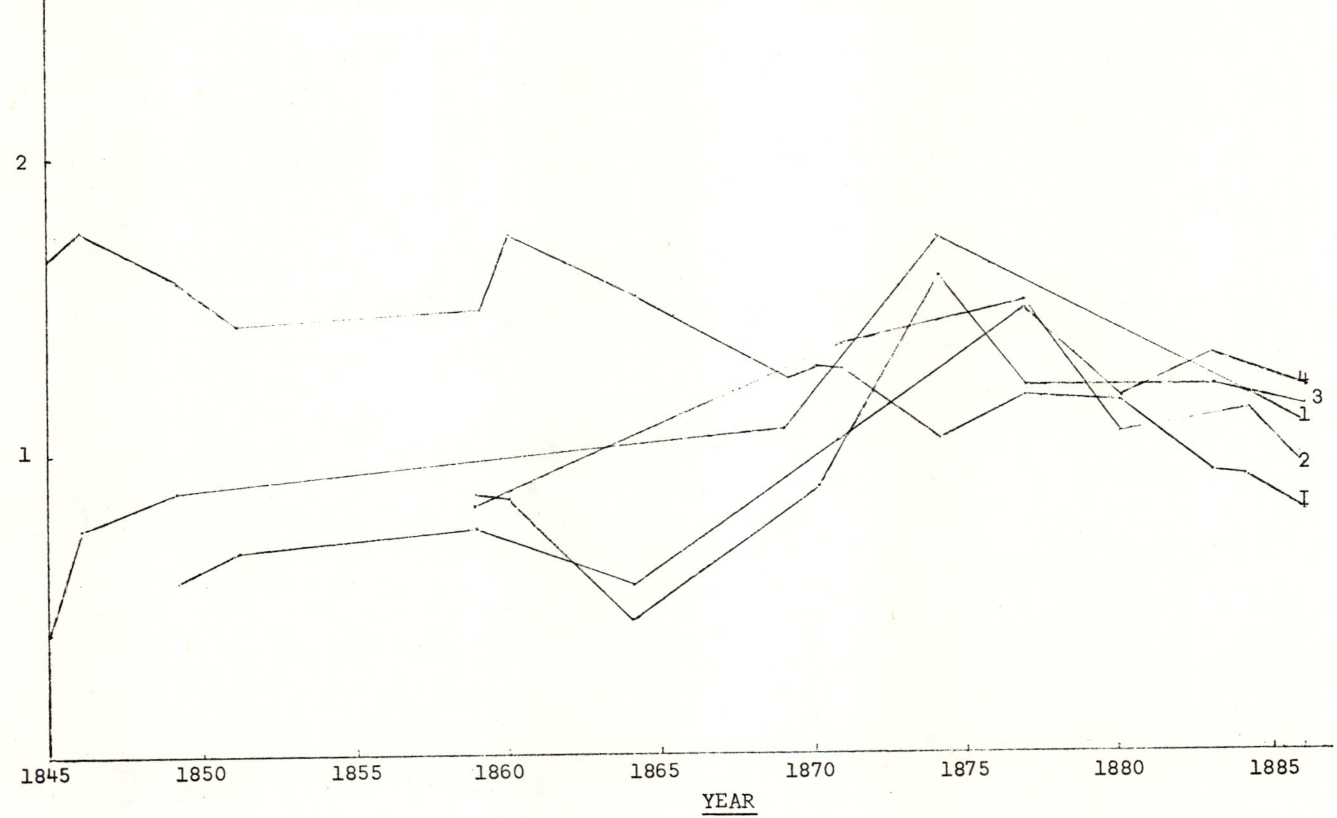

LEGEND:

1. Machinists
2. Millwrights
3. Boilermakers
4. Patternmakers
I. American-British Interest Rate Ratios

The examination of these tables reveal that "real-wages" (or surrogates for the MPP of labor) were lower in the U.S. until approximately 1871 in all occupations. This result is also observed when "real-wages" are measured in terms of pig-iron per worker. It is also clear that only after 1871 did American-British MPP-ratios exceed American-British interest-rate-ratios in all occupations (see Chart 4). The result is repeated when prices of pig-iron are used as deflators. This simply means that the conclusions reached for the previous sectors must be reversed here for the period up to 1871.[48]

E. Summary

At this stage we would like to summarize the main empirical results and their implications, so that we may proceed to the final chapters of this part. We shall combine the results of this chapter and those in the Appendix.

The evaluation would have been much easier and the implications could have been much stronger if for all surrogates in all sectors, one factor was relatively "cheaper" in one country than in the other. The comparisons from our series show some uniformity of the results. Nevertheless, one can still make the following conclusions for each sector:

---

48. This result may conceivably be due to our simplifying assumptions with respect to the interpretation of the statistical data. However, it may not be in complete contrast with what one might expect from a rapidly growing American iron industry (especially after 1870).

a. The cotton industry. If surrogates for "w" and "r" are in terms of final manufacturing output and long term interest rates respectively, then the results generally imply that the American capital-intensity in this sector was higher throughout the period in question. This result is stronger if surrogates for "r" remained unchanged but those for "w" are now money-wage-rates deflated by the cost of living index.[49] If surrogates for "r" are short term interest rates, one may still conclude that the former implication is non-rejectable, especially when one takes the second type of surrogates for "w".

Also in all occupations and for both types of surrogates, "w" was higher in the U.S.

b. The wool industry. As in the former case, the results confirm or at least do not reject the contention that in this sector capital-intensity was lower in the U.K. Here also "w" was higher in the U.S. throughout the period and for all occupations.

c. The iron and steel industry. Here, if surrogates for "w" and "r" are in terms of final manufacturing outputs and long term interest rates respectively, then the results

---

49. The dollar is used as common currency unit. We converted British wages by means of prevailing exchange rate. However, this raises the possibility of error, since the exchange rate used (though based on the gold standard) did not necessarily reflect the "true" relative purchasing power of the domestic currencies. See further Note (a) for Table A3 in the Appendix.

imply that the American capital-intensity in this sector was lower until roughly 1870. Thereafter, the trend was reversed so that the British sector was then under lower capital-intensity. If, however, surrogates for "w" are money-wages deflated by the cost of living index, then the implication would be that throughout the period, capital-intensity was higher in the U.S. regardless of what surrogate we use for "r". We also note that this implication will not hold if surrogates for "r" are short term interest rates and those for "w" are final manufacturing product.

We can also see that unlike the former cases, here "real-wages" in terms of final manufacturing goods were lower in the U.S. until approximately 1870.

One may ask now whether some indications exist to either support or deny these implications for actual capital-intensity. In the following two chapters we shall try to tackle this question. Again, because of the lack of data, we shall not be able to cover all of the three sectors, but rather concentrate only on the sectors of cotton and wool.

CHAPTER 3

The Actual Capital-Intensity: An Indication for the Cotton and Wool Branches

The former chapters define capital-intensity in terms of physical units, namely, capital-labor proportions. Therefore, we would like to follow this definition in this chapter as well. The question is how to approximate this notion empirically.

The indicator for capital-intensity that we were able to construct is "horsepower-per-worker".[50] Thus we took annual averages of horsepower used and workers employed as surrogates for physical quantities of capital and labor respectively.

For the purpose of this chapter, one must retain the provisional assumption that the quality of labor was the same in both countries for each sector at any given period of time.[51]

The period covered is 1870-1900 for the U.S. and 1850-1870 for the U.K. These series are given in Tables 15-18 as follows:

---

50. It is assumed that one unit of horsepower represents the same amount of real capital in both countries. J.B. Heath uses horsepower per operative as surrogate for capital-intensity in Canada and Britain for the year 1948 (see J.B. Heath, "British-Canadian Industrial Productivity", EJ, December, 1957). Heath also elaborates further on the shortcomings of this surrogate (see Heath, *ibid.*,p. 676).

51. This is indeed a strong assumption. In the second part we shall relax this assumption and subsequently measure labor-augmenting technological progress.

TABLE 15: Relative Use of Horsepower in the Cotton Industry in the U. S. (1870-1900)[52]

Year	Annual Av. Number of Workers(1)	Annual Av. Horsepower Used (2)	Horsepower per Worker (3)
1870	135,369	146,040	1.079
1880	174,659	275,504	1.577
1890	218,876	464,881	2.124
1900	297,929	811,347	2.723

---

52. The sources are as follows:

Column (1): U.S. Census Office, Census Reports Twelfth Census of the U.S. Manufactures: Textiles. Government Printing Office, Washington, D.C., 1902, p. 27.

Column (2): U.S. Census Office, Census Reports Vol. VII Twelfth Census of the U.S. 1900, Manufactures Part 1. Government Printing Office, Washington, D.C., 1902.

TABLE 16: Relative Use of Horsepower in the Cotton Industry in the U. K. (1850-1870)[53]

Year	Annual Av. Number of Workers(1)	Annual Av. Horsepower Used (2)	Horsepower per Worker (3)
1850	570,000	82,000	0.144
1860	646,000	293,000	0.453
1870	680,000	307,000	0.451

---

53. The sources are as follows:

Column (1): The estimate for 1850 is from Sir Arthur Balfour, Chairman, Committee on Industry and Trade, "Survey of Textile Industries: Cotton, Wool, Artificial Silk" London, 1928, p. 143. The estimates for 1860-70 are based on Ellison, T., The Cotton Trade of Great Britain (1st Edition), London: 1886, op.cit., p. 66.

Column (2): Mitchell and Deane, Abstract of British Historical Statistics, London: Cambridge University Press, 1962.

TABLE 17: Relative Use of Horsepower in the Woolen and Worsted Goods Industry in the U. S. (1870-1900)[54]

Year	Annual Av. Number of Workers(1)	Annual Av. Horsepower Used (2)	Horsepower per Worker (3)
1870	92,973	93,117	1.001
1880	105,307	122,944	1.167
1890	119,893	176,612	1.498
1900	125,901	237,028	1.883

54. The sources are as follows:

Column (1): U.S. Census Office, Census Reports, Twelfth Census of the U.S. 1900, Manufactures: Textiles, Washington, D.C., Government Printing Office, 1902.
Column (2): See Note (52).

TABLE 18: Relative Use of Horsepower in the Woolen and Worsted Goods Industry in the U. K. (1850-1870)[55]

Year	Annual Av. Number of Workers(1)	Annual Av. Horsepower Used (2)	Horsepower per Worker (3)
1850	254,000	33,000	0.130
1860	270,000	64,000	0.237
1870	263,000	115,000	0.437

---

55. The sources are as follows:

Column (1): The estimates for 1850 and 1870 are from Balfour, *Survey of Textiles Industries: Cotton, Wool, Artificial Silk*. *op.cit.*, p. 271; and the estimate for 1860 is from Ellison, *The Cotton Trade of Great Britain (1st Edition)*, London: 1886, *op.cit.*, p. 127.

Column (2): Mitchell and Deane, *Abstract of British Historical Statistics*, London: Cambridge University Press, 1962. *op.cit.*, p. 198.

The data for horsepower in the U.S. is compiled by the Bureau of Census and it is not available for years earlier than 1870. The data for the British series which is from the Sessional Papers of the British Parliament does not extend beyond 1870. Thus, there is only one year, 1870, where common comparison of "capital-intensity" is possible. The series of horsepower-per-worker for this year clearly indicate that capital-intensity in both sectors of the U.S. was much higher than in the U.K. Although lack of data does not allow us more comparisons of this type, one may still be expected to conclude that because of similar trends of growth in relative use of horsepower in both countries, capital-intensity was higher in the U.S. in both sectors throughout the period of study.

This result is indeed in accord with our previous implications with respect to cotton and wool. In other words, our indicator shows that capital-intensity was lower, or at least not higher in the U.K. in both sectors during most of the second half of the nineteenth century.

CHAPTER 4

Factor Efficiency and Relative Utilization of Natural Resources: A Surrogate for the Cotton and Wool Sectors

We have seen earlier that "factor prices" were higher in the American textile branches for both labor and capital. This was indicated by all of the surrogates throughout the periods of study. We also said that theoretically both factor-prices can be higher in one country only if the index α_M is higher there. In other words, in the context of our model, superior efficiency is possible if both factors are associated with a relatively higher amount of natural resources (as the result of a relatively higher potential-richness). So the question here is whether under the constraint of available historical data we can construct and subsequently show that an index of this type was higher in the U.S.

We took as surrogate for α_M the degree of utilization of natural resources, raw material per unit of factor. Although this is indeed a rough proxy, one may still expect that a higher degree of resource utilization will result in a relatively higher use of raw materials in production.[56]

---

56. Since raw materials can be imported from other countries (i.e., utilization of others' natural resources) one can expect — in the context of our argument and apart from relative cost analysis — that the higher the share of imports in one country's manufacturing sector (e.g., cotton in the U.K.), the lower would be its α_M. We shall see later that the resulting surrogates for α_M in the cotton sector in the U.K. were indeed considerably lower than those in the U.S. Here, however, one must note that these surrogates are quite "rough" and hence somewhat weak in representing the true concept of α_M.

We used annual averages of workers employed as surrogate for labor, and spindles and looms as two alternatives for capital. Raw cotton and raw wool consumed, which are the main raw materials in production, are used as surrogates for natural resources in the cotton and wool sectors respectively. The period of study covers the second half of the nineteenth century.

One must still retain the previous assumption about the quality of factors in order to make comparisons between the two countries meaningful.

Tables 19-22 give these series as follows:

TABLE 19: Relative Use of Raw Material in the Cotton Industry in the U. S. (1850-1900)[57]

Year	Raw Cotton Consumed 1000 lb.(1)	Annual Averages of Factors Employed			Raw Cotton per - (lb.)		
		Workers (2)	Spindles 1000's(3)	Looms (4)	Worker (5)	Spindle (6)	Loom (7)
1850	272,527	92,286	3,634	--	2953.07	74.99	--
1860	422,705	112,028	5,236	126,313	3773.20	80.73	3346.49
1870	398,308	135,369	7,132	157,310	2942.39	55.84	2531.99
1880	750,344	174,659	10,653	225,759	4296.05	70.43	3323.65
1890	1,117,946	218,876	14,188	324,866	5107.67	78.79	3441.25
1900	1,814,003	297,929	19,051	450,682	6088.71	95.22	4025.02

57. The sources are as follows:

Column (1): U.S. Census Office, *Census Reports, Twelfth Census of the U.S. 1900, Manufactures: Textiles*, Washington, D.C.:Government Printing Office, 1902. *op.cit.*, p. 27. *For 1850* we took from U.S. Census Office, *Manufactures of the U.S. in 1860, Eighth Census*, Washington, D.C., Government Printing Office, 1865.

Column (2): See Note (52).

Columns (3)-(4): U.S. Census Office, *Census Reports, Twelfth Census of the U.S. 1900, Manufactures: Textiles*, *op. cit.*, p. 55.

TABLE 20: Relative Use of Raw Material in the Cotton Industry in the U. K. (1845-1890)[58]

Year	Raw Cotton Consumed 1000 lb.(1)	Annual Averages of Factors Employed			Raw Cotton per - (lb.)		
		Workers (2)	Spindles 1000's(3)	Looms (4)	Worker (5)	Spindle (6)	Loom (7)
1845	607,000	540,000	19,500	282,000	1124.07	31.13	2152.47
1850	588,000	570,000	21,977	250,000	1031.58	26.76	2352.00
1860	1,084,000	646,000	30,400	400,000	1678.02	35.66	2710.00
1870	1,078,000	680,000	37,719	441,000	1585.30	28.58	2444.44
1880	1,361,000	686,000	42,000	550,000	1983.96	32.40	2474.55
1890	1,664,000	740,000	45,000	660,000	2248.65	36.98	2521.21

58. The sources are as follows:

Column (1): Mitchell and Deane, _Abstract of British Historical Statistics_, _op.cit._, p. 179.
Column (2): See Note (53).
Column (3): Ellison, _The Cotton Trade of Great Britain_, _op. cit._, p. 68. The estimates for 1870 and 1890 are from Mitchell and Deane, _Abstract of British Historical Statistics_, _op.cit._, p. 185.
Column (4): Ellison, _The Cotton Trade of Great Britain_, _op.cit._, p. 69. The estimates for 1850, 1870 and 1890 are from Mitchell and Dean, _Abstract of British Historical Statistics_, _op.cit._, p. 185.

TABLE 21: Relative Use of Raw Material in the Woolen and Worsted Goods Industry in the U. S. (1850-1900)[59]

Year	Raw Wool Consumed 1000 lb.(1)	Annual Averages of Factors Employed			Raw Wool per - (lb.)		
		Workers (2)	Spindles 1000's(3)	Looms (4)	Worker (5)	Spindle (6)	Loom (7)
1850	62,382	40,000	--	--	1559.55	--	--
1860	82,804	43,738	640	16,075	1893.18	129.44	5151.10
1870	206,842	92,973	2,046	40,311	2224.75	101.09	5131.16
1880	274,832	105,307	1,997	48,594	2609.82	137.63	5655.68
1890	337,479	119,893	2,570	58,213	2814.83	131.29	5797.31
1900	367,142	125,901	3,278	61,253	2916.11	112.01	5993.86

---

59. The sources are as follows:

Column (1): Estimates based on numbers given in U.S. Census Office, _Census Reports, Twelfth Census of the U.S. 1900, Manufactures: Textiles_, _op.cit._, p. 123.
Column (2): See Note (53).
Columns (3)-(4): U.S. Census Office, _Census Reports, Twelfth Census of the U.S. 1900, Manufactures: Textiles_, _ibid._, p. 99 and p. 102.

TABLE 22: Relative Use of Raw Material in the Woolen and Worsted Goods Industry in the U. K.

(1850-1900)[60]

Year	Raw Wool Consumed 1000 lb. (1)	Annual Averages of Factors Employed			Raw Wool per - (lb.)		
		Workers (2)	Spindles 1000's (3)	Looms (4)	Worker (5)	Spindle (6)	Loom (7)
1850	241,000	254,000	2,471	42,056	948.82	97.53	5730.45
1860	312,300	270,000	3,472	64,818	1156.67	89.95	4818.11
1870	453,500	263,000	4,524	112,272	1724.33	100.24	4039.29
1880	496,500	316,000	5,685	142,386	1571.20	86.28	3486.93
1890	622,200	302,000	5,851	130,271	2060.26	106.34	4776.20
1900	655,600	260,000	5,845	117,392	2521.54	112.16	5584.33

---

60. The sources are as follows:

Column (1): Balfour, Survey of Textiles Industries: Cotton, Wool, Artificial Silk, op.cit., p. 275.
Column (2): See Note (54).
Columns (3)-(4): Deane, P. and W.A. Cole, British Economic Growth 1688-1959: Trends and Structure, London: Cambridge University Press, 1962. p. 200.

It is clear from these Tables that except for one year, 1900, where raw-wool-per-spindle was the same in both countries, the rest strongly indicate higher ratios in the U.S. All of the three proxies — raw-material-per-worker, per-spindle, and per-loom — were higher in the U.S. in both sectors. This result makes the conclusion somewhat stronger because none of these ratios were lower in the U.S., namely, there was strict inequality throughout the period and for all cases (except for the above case).

Granting that these are "tolerable" surrogates for the index of α_M, then this result is indeed the one we expected to get. In other words, the contention that both factors were more "expensive" in the U.S. (or equivalently in our model — both were more efficient) imply that they were associated with a higher α_M . But this was clearly indicated by our surrogates in both sectors.

CHAPTER 5

Conclusion

At this stage, we would like to summarize and perhaps briefly evaluate the results reached so far. First, we would like to emphasize that the former chapters did not attempt to present an alternative explanation to the problem of American industrial efficiency as compared with the British (under the assumption of identical technologies). Instead, the problem was first postulated theoretically and subsequently, historical evidence was used to show whether the indication was in favor of the traditional view of Rothbarth and Habakkuk. The indicators for actual capital-intensity show that for at least two manufacturing sectors — cotton and wool — one may accept the contention that the Americans used higher capital-intensity in production. The case of the third large manufacturing sector — iron and steel — suffered from the lack of data needed to derive the same evidence. The structure of factor-prices in the latter case, however, did not imply conclusive results, so further inquiry is needed here.

In the case of textile branches, the last chapter indicated that possibly more natural resources were utilized and associated with each unit of factor which may explain the fact that both "real-wages" and interest rates were higher in the U.S. during our period of study. These results were in accord with the theoretical implications of our model.

Our empirical study is based on a simplified theory, and so like any other theoretical models, it may involve specification error. Our conclusions are based on comparative statics. It was also assumed that both systems were competitive, the same technology was simultaneously available to them and that factors were homogeneous in each and for both economies. This clearly is not the true picture of the world as it is. It may be false to assume that both countries were using only two types of factors. Factors may not have been homogeneous (and neither is output). Technology may have been different in each country. Also, depreciation was clearly accounted for in both countries.[61] It would be an injustice, however,

---

61. The abstraction from depreciation enabled us to treat interest-rate ratios as relative MPP of capital. In other words, interest rates were considered as surrogates for both gross real-rental and net real-rental of capital. In the absence of allowance for depreciation, equilibrium interest rate equals MPP of capital (or gross real-rental). Obviously, once depreciation is introduced into the picture, the equality between interest rate and MPP of capital will be violated. In equilibrium, MPP of capital will exceed interest rate by the amount of the depreciation rate. Thus, interest-rate ratios may become a biased estimate for the relative MPP of capital. The direction of this bias will depend on the relative rate of depreciation in each country. To illustrate this point more rigorously, we may follow Samuelson's approach (see P.A. Samuelson, "Equalization by Trade of the Interest Rate Along with the Real Wage", in Essays in Honor of Gottfried Haberler, <u>Trade, Growth and the Balance of Payments</u>, Chicago, 1965). Assume that depreciation is a constant percentage per unit of time regardless of age. Let us denote this rate by m. At equilibrium we have:

$$R_g^i = P_M^i \frac{\partial X_M^i}{\partial K} = P_M^i \alpha_M^i \frac{df^i}{dk} \quad ,$$

$$r_g^i = \frac{R_g^i}{P_M^i} \quad ,$$

$$R_n^i = R_g^i - m^i P_M^i$$

$$r^i \equiv r_n^i = \frac{R_n^i}{Pi_M} = \alpha_M^i \frac{df^i}{dk} - m^i \quad , \qquad i=1,2;$$

where:

R_g = gross rental of capital

r_g = gross real rental of capital

R_n = net rental of capital

r, r_n = net real rental of capital (or interest rate)

P_M = price of the capital good (or of the manufacturing good)

It follows that:

$$\rho^i = r^i + m^i \qquad\qquad i=1,2;$$

where

$$\rho \equiv \frac{\partial X_M}{\partial K}$$

Given this formulation, one should amend relationships (5) (in chapter 1 of Part I) as follows:

$$(5) \qquad \frac{w^1}{\rho^1} \gtrless \frac{w^2}{\rho^2} \qquad\qquad k^1 \gtrless k^2$$

It follows that our empirical observations

$$\frac{w^1}{w^2} \geq \frac{r^1}{r^2} \quad (>1)$$

in the case of cotton, wool, and iron and steel (after, approximately, 1870) imply higher capital intensity in the U.S. only if

$$\frac{r^1}{r^2} \geq \frac{\rho^1}{\rho^2} \quad ;$$

namely, only if the interest-rate ratio, $\frac{r^1}{r^2}$, is an over-estimate of the relative MPP of capital, $\frac{\rho^1}{\rho^2}$. Because then we have

$$(\frac{w^1}{w^2} > \frac{r^1}{r^2} \qquad [(\frac{w^1}{w^2} > \frac{\rho^1}{\rho^2}) \qquad (k^1 > k^2)].$$

The question is under what conditions $\frac{r^1}{r^2}$ will be an over-estimate for $\frac{\rho^1}{\rho^2}$. The answer to this question clearly depends on the relationship between m^1 and m^2. Our results will not be violated (after the introduction of depreciation) only if

$$w > \frac{\rho^1}{\rho^2} , \qquad \text{where } w \equiv \frac{w^1}{w^2} > 1.$$

namely, only if

$$w > \frac{r^1 + m^1}{r^2 + m^2}$$

or

$$m^1 < Wr^2 + Wm^2 - r^1$$

$$\text{(i)} \quad \therefore\ m^1 < A + Wm^2 \ ;$$

where

$$A \equiv Wr^2 - r^1 \geqq 0, \text{ (since } m^i \geq 0,\ i=1,2).$$

In the last inequality we know the magnitudes of A and W. Thus we can define the relationship between m^1 and m^2 for which our results will still hold. This can be done, simply, by drawing the curve corresponding to $m^1 = A + Wm^2$. All the combinations of m^1 and m^2 which lie beneath the curve will be consistent with our findings. Diagram C illustrates these conditions as follows:

<u>Diagram C</u>: The Relationship Between Rates of Depreciation and the Consistency of the Empirical Results

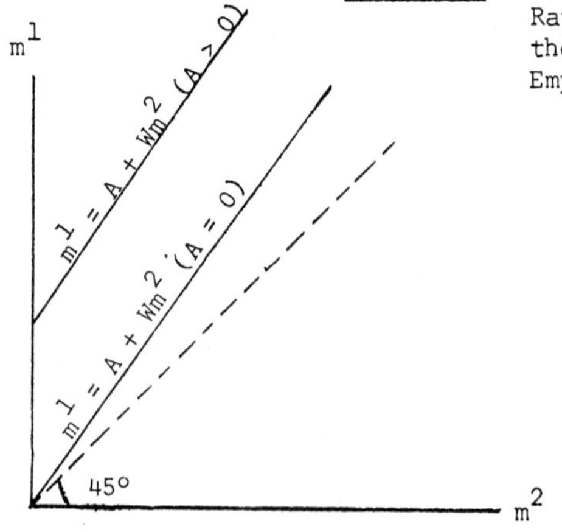

It is clear from this diagram that our results will hold for all combinations of m^1 and m^2 such that $m^1 \leq m^2$ (i.e., all combinations on the 45° line or beneath it). This is clearly a sufficient but not necessary condition. It can also be readily seen that $m^1 > m^2$ is a necessary condition for our results to be violated. However, this is not a sufficient condition. The depreciation rate can be higher in the U.S. and one can still conclude that

$$\left(\frac{w^1}{w^2} > \frac{r^1}{r^2}\right) \Longrightarrow \left(\frac{w^1}{w^2} > \frac{\rho^1}{\rho^2}\right), \qquad \text{hence}$$

retaining our previous results. (Empirically, no one tried to estimate the magnitudes of m's for the aggregate manufacturing sectors in both countries. Paul F. McGouldrick estimates the rate of depreciation for the New England textiles during 1836-1886 at 1.75-3.00% annually (see McGouldrick, Paul F., New England Textiles in the Nineteenth Century, Harvard University Press, Cambridge, Massachusetts, 1968, p. 69.) He does not give, however, international comparisons. Habakkuk claims that relative scarcity of labor in the U.S. has induced "greater readiness to scrap existing equipment in order to take advantage of technical progress" (see Habakkuk, American and British Technology in the Nineteenth Century: The Search for Labor-Saving Inventions, op.cit., p. 62). He also asserts that American machines and equipment were relatively more flimsy and short-lived as compared with the British (ibid., pp. 85-90). This suggests a higher rate of depreciation in the U.S. If this was indeed the fact, our results become somehow weaker. However, as was indicated, this possibility may not be in contrast with our previous findings.

to dismiss all of these assumptions. There is one assumption which could indeed find a firm base in the economic organization of the nineteenth century, namely, the assumption of competitive systems. In both of the American and British economic organizations, the forces of free markets played a predominant role. Consumers and producers were to a greater extent price-takers in both systems. Nevertheless, specification error may still exist. However, one may commit a greater error by disregarding the whole question altogether.

Part II

Relative Productivity and Factor-Augmenting Technological Progress Under Different Technologies

In the previous part, the question of relative productivity was examined under the assumption that technologies were the same in both countries. The definition of "the same technology" was simply that the elasticity of substitution is the same for any given capital-labor ratio. In this part, we define "different technologies" by the same criterion, namely, that each country has different elasticity of substitution (for any given capital-labor ratio) in each sector. We shall assume, however, that in both countries, the elasticities of substitution are constant. In other words, we shall assume production functions with constant but different elasticities in both countries.[62]

As we have indicated in the introduction, we shall empirically examine the implications of Habakkuk's basic presumption about relative factor-scarcity. Habakkuk's conclusion was that relative-dearness of labor in the U.S. has resulted in labor-saving technological progress and hence, in a higher relative labor efficiency in this country. In the U.K., on the other hand, one might have expected capital-saving bias in technology as a result of relative capital-scarcity. The main purpose of this part, then, will be to try to measure the rate of efficiency growth in both countries and see whether indeed there is an indication to

---

62. Diagrammatically, the slope of isoquants for any sector is given by constant elasticity of substitution. The shape, however, differs between the two countries (see later section D in the Appendix.)

support Habakkuk's implications.

We shall relate the concept of biased technological change toward either relative labor-augmentation or relative capital-augmentation to the Hicksian notions of neutral, labor-saving and capital-saving technological progress. According to Hicks,[63] inventions are to be classified as neutral, labor-saving or capital-saving, according to whether they leave unchanged the marginal productivity of labor relative to the marginal productivity of capital, lower the relative marginal productivity or increase the marginal productivity of labor relative to that of capital -- given a constant capital-labor ratio. These definitions will be adopted in the current approach.

Under the constraint of available data, we shall be examining the cases of cotton and wool in both countries. First, however, we shall draw the theoretical basis for this study.

---

63. Hicks, J.R., The Theory of Wages (1st Edition), London: MacMillan, 1932, p. 15.

CHAPTER 6

The Model

The following model is the now-familiar constant returns to scale production function with constant elasticity of substitution. This model was applied by David and Van Klundert[64] in their study of the U.S. private domestic economy. Since we shall follow their approach very closely, we shall stress here very briefly the main characteristics of the model, leaving the rest of the theory in the Appendix (see section D there).

We assume that each of the manufacturing sectors has the following form of production function:

$$(1)\qquad V = [(E_L L)^{-\rho} + (E_K K)^{-\rho}]^{-1/\rho}$$

Each country's sector is assumed to have different elasticity of substitution, σ, and hence different ρ, (see definitions in the Appendix section D).

We interpret changes in E_L and E_K over time as labor-augmenting and capital-augmenting technological changes. We note, however, that this does not say anything about the sources of such changes in efficiency growth. An increase in labor efficiency or labor-augmenting change means here an increase in E_L. This can be done, for example, by the introduction of better machines. On the other hand, an increase in E_K is designated here

---

64. David, Paul A. and Th. Van Klundert, "Biased Efficiency Growth and Capital-Labor Substitution in the U.S. 1899-1960", AER, June 1965.

as capital-associated efficiency growth or capital-augmenting technological change and can be a consequence of improved labor qualitites.

As it was stated above, the notion of factor-augmenting technical change and its biases is related to the Hicksian definitions. The relationship between factor-augmenting technological change and the Hicksian definitions of factor-saving innovations is stated in equation (9a) as follows:

$$\text{(9a)} \quad \frac{\dot{m}_L}{m_L} - \frac{\dot{m}_K}{m_K} = \frac{\sigma-1}{\sigma}\left(\frac{\dot{E}_L}{E_L} - \frac{\dot{E}_K}{E_K}\right).$$

It is clear from this expression that Hicksian neutral technological change can be expressed in our model by the equality of the rates of growth of labor and capital efficiency, namely, when $\dot{E}_L/E_L = \dot{E}_K/E_K$.

We can also see that a labor-augmenting bias in technical change, i.e., when $\dot{E}_L/E_L > \dot{E}_K/E_K$, is equivalent to a labor-saving innovation in the Hicksian sense, if and only iff the elasticity of substitution, σ, is <u>less than unity</u>.

Similarly, what we designate as capital-augmenting bias, i.e., when $\dot{E}_K/E_K > \dot{E}_L/E_L$, amounts to the same thing as a capital-saving innovation in the Hicksian sense, only if the elasticity of substitution is <u>less than unity</u>.

It is also obvious that labor-augmenting bias in our model can be interpreted as Hicksian capital-saving technological change when the elasticity of substitution is <u>greater than unity</u>.

We shall make use of these relationships when we come to evaluate our empirical results later on.

The regression model that we shall use is basically the one given by equation (13) (in section D of the Appendix). However, we shall follow first the David and Van Klundert approach and use their regression model which incorporates lags in the adjustments of both of the actual capital-intensity and the relative factor-shares to their long term desired levels. This was given in their expression (20.)[65] Our data refers to factors actually employed, so we shall omit the reference to unemployment. The least-squares regression model becomes then as follows:

$$(16)\quad \ln\left(\frac{K}{L}\right)_t = c_o + \alpha\left(\frac{\sigma}{1-\sigma}\right) \ln \left(\frac{\pi_L}{\pi_K}\right)_t + \alpha(\lambda_L - \lambda_K)_t + (1-\alpha)\ln\left(\frac{K}{L}\right)_{t-1};$$

where

$\alpha \equiv$ elasticity of adjustment indicating the fraction of the desired adjustment that is completed in the course of a single year.

The restrictive nature of this model is that biased factor-augmenting technical change either labor-saving or capital, can only proceed at a steady rate through time. This assumption, however, is vital to the identification of the production function. As Nerlove[66] points out, in order to avoid the Diamond-McFadden

---

65. See David and Van Klundert, *ibid.*, p. 376.

66. Nerlove, Marc, "Recent Empirical Studies of the CES and Related Production Functions", in M. Brown (ed.) *The Theory and Empirical Analysis of Production Studies in Income and Wealth* Vol. 31. National Bureau of Economic Research, New York, 1967.

impossibility theorem, one must assume that all technical change is factor-augmenting and exponential.[67] This was assumed in our approach.

Next we shall apply this model to the textiles sectors in both countries to estimate the parameters σ and $(\lambda_L - \lambda_K)$. The estimation of these parameters will enable us, of course, to transform the results into the Hicksian definitions of factor-saving technological inventions, and hence we may evaluate accordingly the Habakkuk contention.

Data was available for two surrogates for K. Thus, we shall have two estimates for each of these parameters — each is the result of a different capital-labor ratio. One should not, however, expect identical estimates. In fact, the resulting estimates are not identical, but they can still give rough estimates for each of σ and $(\lambda_L - \lambda_K)$ and hence form a basis for our forthcoming evaluation.

In anticipation, we might add that in all but one of the cases, the resulting estimates are consistent with the basic assumptions of the model. Only in one case — wool in the U.K. — one of the two estimates for σ was negative. We shall disregard

---

67. In essence, the Diamond-McFadden "Impossibility" theorem states that it is impossible to measure either the bias or the elasticity of substitution with a neoclassical production function. They state that series of observable variables can be generated by more than one production function having an arbitrary elasticity or arbitrary bias. Nerlove shows, on the other hand, that identification becomes possible only if one assumes that all of the technical change is biased, factor-augmenting, and exponential. See Nerlove, ibid., pp. 92-99.

this case, treating it not as a case of complete specification error, but rather as a possible result of the relatively small number of observations taken for this case.

CHAPTER 7

Estimation of Biased Factor-Augmenting Technological Growth in the Cotton Industry in the U.S. and the U.K.

In fitting the available data to our regression model, we took as proxy for labor annual averages of workers employed. As surrogate for capital, we took annual averages of machines of a certain type which were used in the industry. The latter are indeed closer to the concept of K as it is defined in the model.[68] Data, however, was available for two types of machines in both countries, namely, spindles and looms. Thus we have constructed two types of capital-labor-ratios — spindles per worker and looms per worker. We have used both of these series as the dependent variable in the regression model.

The period of study covers the years 1850-1900 for the U.S. and the period 1820-1880 for the British sector. The number of observations, however, was quite limited because the data was available only decennially.

Labor share, π_L, was calculated as a ratio between the annual wage-bill and the annual value-added for the industry. The capital share, π_K, was calculated as $(1-\pi_L)$ (see section E and Tables A21-A24 in the Appendix.)

This data is summarized in Tables 23-24 as follows:

---

68. In most of the empirical studies, annual value of stock of capital is used as proxy for K with some adjustment for unemployed capital.

TABLE 23: Factor Proportions and Factor Shares in the Cotton Industry in the U. S. (1850-1900)[69]

Year	k_1 (1)	k_2 (2)	Π_L (3)	$\Pi_K = 1 - \Pi_L$ (4)	Π_L / Π_K (5)
1850	39.37	1.04	0.566	0.434	1.304
1860	46.70	1.04	0.373	0.627	0.595
1870	52.68	1.16	0.508	0.492	1.033
1880	60.99	1.29	0.400	0.600	0.667
1890	64.80	1.48	0.438	0.562	0.779
1900	63.94	1.51	0.405	0.595	0.681

---

69. k_1 and k_2 stand for spindles-per-worker and looms-per-worker respectively. The sources for these series are given in Note (57). The series for the π's and their sources are given in Table A21 in the Appendix.

TABLE 24: Factor Proportions and Factor Shares in the Cotton Industry in the U. K. (1820-1880)[70]

Year	k_1 (1)	k_2 (2)	Π_L (3)	$\Pi_K=1-\Pi_L$ (4)	Π_L/Π_K (5)
1820	15.73	0.57	0.449	0.551	0.815
1830	18.87	0.57	0.458	0.542	0.845
1840	30.94	0.55	0.431	0.569	0.757
1850	38.56	0.44	0.415	0.585	0.709
1860	47.05	0.62	0.406	0.594	0.683
1870	55.44	0.65	0.418	0.582	0.722
1880	61.22	0.80	0.428	0.572	0.748

---

70. The definition of k_1 and k_2 is the same as in the previous case. The sources for these series are given in Note (58). The series for the π's and their sources are given in Table A22 in the Appendix.

The results of fitting equation (16) to observations on k_1, k_2 and $\pi_L/1-\pi_L$ for the U.S. cotton industry during the period 1850-1900 by simple least-squares are shown by (Ia) and (Ib) as follows:

$$\text{(Ia)}\quad \ln(k_1)_t = 3.6600 + \underset{(0.1606)}{0.0477} \ln \left(\frac{\pi_L}{1-\pi_L}\right)_t + $$

$$+ 0.0078t + 0.0434 \ln(k_1)_{t-10} \; ;$$

and with

$$R = 0.9757,$$

$$\text{Std. error est.} = 0.0435$$

$$\text{Degrees of freedom} = 2.$$

(The numbers in parentheses are standard-error of the correlation coefficients).

Making use of the definitions of the coefficients in (16) we obtain the following parameter estimates:

$$\hat{\alpha} = 0.9566$$

$$\widehat{\lambda_L - \lambda_K} = 0.0082$$

$$\hat{\sigma} = 0.0475$$

$$\text{(Ib)}\quad \ln(k_2)_t = 0.0057 + \underset{(0.0769)}{0.0898} \ln \left(\frac{\pi_L}{1-\pi_L}\right)_t + $$

$$+ \underset{(0.0035)}{0.0082t} + \underset{(0.4205)}{0.1530} \ln(k_2)_{t-10} \; ;$$

and with

$$R = 0.9857,$$

$$\text{Std. error est.} = 0.0280,$$

$$\text{Degrees of freedom} = 2.$$

Here we obtain the following parameter estimates:

$$\hat{\alpha} = 0.8470,$$

$$\widehat{\lambda_L - \lambda_K} = 0.0097,$$

$$\hat{\sigma} = 0.0959.$$

Similarly, the results of fitting the same formula to observations on k_1, k_2 and $(\pi_L/1-\pi_L)$ for the U.K. cotton industry during the period 1820-1880 yields (IIa) and (IIb) as follows:

$$\text{(IIa)} \quad \ln(k_1)_t = 2.4066 - \underset{(0.4752)}{1.7207} \ln\left(\frac{\pi_L}{1-\pi_L}\right)_t + \underset{(0.0024)}{0.0177t} + \underset{(0.0347)}{0.0369} \ln(k_1)_{t-10};$$

and with

$$R = 0.9964,$$

$$\text{Std. error est.} = 0.0449,$$

$$\text{Degrees of freedom} = 3.$$

From this we obtain the following parameter estimates:

$$\hat{\alpha} = 0.9631,$$

$$\widehat{\lambda_L - \lambda_K} = 0.0184,$$

$$\hat{\sigma} = 2.2716.$$

$$\text{(IIb)} \quad \ln(k_2)_t = -0.2941 + \underset{(1.2087)}{1.5849} \ln\left(\frac{\pi_L}{1-\pi_L}\right)_t + \underset{(0.0038)}{0.0095t} + \underset{(0.2948)}{0.1283} \ln(k_2)_{t-10};$$

and with

$$R = 0.8258,$$

$$\text{Std. error est.} = 0.1024,$$

$$\text{Degrees of freedom} = 3.$$

The resulting parameter estimates are

$$\hat{\alpha} = 0.8717,$$

$$\widehat{\lambda_L - \lambda_K} = 0.0109,$$

$$\hat{\sigma} = 0.6452.$$

Now let us summarize the results reached thus far. First, we should say that from an econometric point of view, the results are acceptable, but not as strong as one might wish. In all cases the correlation coefficients lead to meaningful parameter estimates. However, in most cases, these coefficients are accompanied by significantly high standard errors. In the case of American cotton and wool industries, at least, this result turned out to be not as severe as it may look. Estimates based on cross-section data for these sectors (for which data was available) surprisingly show similar magnitudes for the estimates of σ with significantly low standard errors (see section G in the Appendix).

From (Ia) and (Ib) we notice that both estimates for $(\widehat{\lambda_L - \lambda_K})$ are positive. This means that there was labor-augmenting technical change in the American cotton industry during this period. It is also clear from these equations that the estimated elasticity of substitution is less than unity. Thus, we may conclude that there was a labor-saving bias in technological growth, defined in the Hicksian sense, in this sector during this period. More specifically, we can see that the estimated annual rate of growth of this bias was approximately 0.8-0.9 of a percentage point. In other words, during this period the annual rate of labor-augmentation has proceeded at approximately 0.8-0.9 of a percentage point higher than the annual rate of capital augmentation.

Now looking at the case of cotton in the U.K., we can see that there is not such a clear trend as in the previous case. From (IIa) and (IIb) it is clear that there was a positive growth rate of relative labor-augmentation during the period 1820-1880. However, in the Hicksian sense, the result is not unambiguous. From (IIa) (i.e., when spindles-per-worker is taken as a proxy for capital-labor-ratio), we obtain an estimation for the elasticity greater than unity. This implies that during this period there was capital-saving bias in technology in the Hicksian terminology. From (IIb) (i.e., when looms-per-worker is taken as a surrogate for the capital-labor ratio), the resulting estimate for the elasticity is less than unity which implies that the bias was labor saving.

Following Habakkuk's argument of relative capital-scarcity in the U.K., one might have expected estimation of capital saving bias in technological change. Our resulting estimates do not completely agree with this contention but still do not reject it completely. So one may still suggest, but not without ambiguity, that the cotton industry in the U.K. was experiencing capital-saving technological growth during the second half of the nineteenth century.

A Digression: Estimation Without Lags

Since our observations are based on decennial data, one may argue that a ten-year period is sufficient enough to allow full adjustment of the variables. Thus it is desirable to estimate our parameters without lagged-adjustments. This can be done, simply, by fitting equation (13) to this data.

The results of fitting this equation to observations on k_2 and $\pi_L/1-\pi_L$ for the American cotton industry are shown by (I_b') as follows:

$$(I_b') \quad \ln(k_2)_t = -0.0017 + \underset{(0.0648)}{0.0906} \ln\left(\frac{\pi_L}{1-\pi_L}\right)_t + \underset{(0.0010)}{0.0094t};$$

and with

$$R = 0.9847,$$

$$\text{Std. error est.} = 0.0289,$$

$$\text{Degrees of freedom} = 3.$$

Thus, we obtain the following parameter estimates:

$$\hat{\sigma} = 0.0831,$$

$$\widehat{\lambda_L - \lambda_K} = 0.0094.$$

Similarly, for the British cotton industry, the results of fitting equation (13) to observations on k_1 and k_2 and $\pi_L/1-\pi_L$ are shown by (II'_a) and (II'_b) as follows:

$$(II'_a) \quad \ln(k_1)_t = 2.4400 - \underset{(0.4771)}{1.7972} \ln\left(\frac{\pi_L}{1-\pi_L}\right)_t + \underset{(0.0017)}{0.0195}t;$$

and with

R = 0.9950,

Std. error est. = 0.0526,

Degrees of freedom = 4.

so that

$\hat{\sigma} = 2.2544,$

$\widehat{\lambda_L - \lambda_K} = 0.0195.$

$$(II'_b) \quad \ln(k_2)_t = -0.2867 + \underset{(0.9571)}{1.8280} \ln\left(\frac{\pi_L}{1-\pi_L}\right)_t + \underset{(0.0034)}{0.0094}t;$$

and with

R = 0.8135,

Std. error est. = 0.1056,

Degrees of freedom = 4.

Thus we obtain

$\hat{\sigma} = 0.6464,$

$\widehat{\lambda_L - \lambda_K} = 0.0094.$

(The case of the American cotton industry with observations on k_1 is omitted because the resulting estimate for σ was not meaningful).

It can be readily seen that the difference between estimates in this section and those in the previous one, which involved lagged-adjustments, is quite negligible. So the conclusions remained the same, thus needing no further elaboration.

CHAPTER 8

Estimation of Biased Factor-Augmenting Technological Growth in the Wool Industry in the U.S. and the U.K.

The nature of the data for this industry is similar to that of the cotton branch. As a surrogate for L we took again annual averages of workers employed. Spindles and looms are taken again as alternative proxies for K; so spindles-per-worker and looms-per-worker define again two alternative approximations to the concept of the capital-labor ratio. The data again is given decennially, but the period for both countries is now 1850-1900.

Factor-shares are defined in the same manner as in the previous case, namely, π_L is calculated as a ratio between the annual wage-bill and the annual net value-added for the wool manufacturing sector, while π_K equals $(1-\pi_L)$.

As we indicated before, we shall present parameter estimates for the U.K. by using only k_2 as the dependent variable. We shall present, however, both series of capital-labor ratios. This data is given in Tables 25-26 as follows:

TABLE 25: Factor Proportions and Factor Shares in the Wool Industry in the U. S. (1850-1900)[71]

Year	k_1 (1)	k_2 (2)	Π_L (3)	$\Pi_K=1-\Pi_L$ (4)	Π_L/Π_K (5)
1850	14.62	0.37	0.472	0.528	0.894
1860	14.62	0.37	0.390	0.610	0.639
1870	22.01	0.43	0.479	0.521	0.919
1880	18.96	0.46	0.458	0.542	0.845
1890	21.44	0.49	0.526	0.474	1.101
1900	26.03	0.49	0.500	0.500	1.000

---

71. The definition of k's is as before. Their sources are given in Note (59). The series for the π's are given in Table A23 in the Appendix.

TABLE 26: Factor Proportions and Factor Shares in the Wool Industry in the U. K. (1850-1900)[72]

Year	k_1 (1)	k_2 (2)	Π_L (3)	$\Pi_K=1-\Pi_L$ (4)	Π_L/Π_K (5)
1850	9.72	0.16	0.295	0.705	0.419
1860	12.85	0.24	0.325	0.675	0.482
1870	17.19	0.42	0.258	0.742	0.347
1880	17.97	0.45	0.373	0.627	0.596
1890	19.37	0.43	0.284	0.716	0.397
1900	22.46	0.45	0.245	0.755	0.324

---

72. The definition of k's is as before. Their sources are given in Note (60). The series for the π's are given in Table A24 in the Appendix.

For the U.S. wool industry, the results of fitting expression (16) to observations on k_1, k_2 and ($\pi_L/1-\pi_L$) by simple least-squares are shown by (IIIa) and (IIIb) as follows:

$$\text{(IIIa)} \quad \ln(k_1)_t = 2.7516 + \underset{(0.6165)}{0.5629} \ln\left(\frac{\pi_L}{1-\pi_L}\right)_t + \underset{(0.0092)}{0.0048t} + \underset{(0.1140)}{0.0618} \ln(k_1)_{t-10};$$

and with

$$R = 0.9246,$$

$$\text{Std. error est.} = 0.0889,$$

$$\text{Degrees of freedom} = 2.$$

The resulting estimates for our parameters are:

$$\hat{\alpha} = 0.9382,$$

$$\widehat{\lambda_L-\lambda_K} = 0.0051,$$

$$\hat{\sigma} = 0.3750.$$

$$\text{(IIIb)} \quad \ln(k_2)_t = -0.9647 + \underset{(0.1472)}{0.2568} \ln\left(\frac{\pi_L}{1-\pi_L}\right)_t + \underset{(0.0016)}{0.0046t} - \underset{(0.0644)}{0.0562} \ln(k_2)_{t-10};$$

and with

$$R = 0.9833,$$

$$\text{Std. error est.} = 0.0236,$$

$$\text{Degrees of freedom} = 2.$$

The resulting parameter estimates are:

$$\hat{\alpha} = 1.0562,$$

$$\widehat{\lambda_L - \lambda_K} = 0.0044,$$

$$\hat{\sigma} = 0.1956.$$

For the British case we obtain the following relationships:

$$\text{(IV)} \quad \ln(k_2)_t = \underset{}{-1.5937} + \underset{(0.6151)}{0.2171} \ln \left(\frac{\pi_L}{1-\pi_L}\right)_t + \underset{(0.0073)}{0.0205t} - \underset{(0.2082)}{0.1829} \ln(k_2)_{t-10};$$

and with

$$R = 0.9070,$$

$$\text{Std. error est.} = 0.1831,$$

$$\text{Degrees of freedom} = 2.$$

The resulting estimates for our parameters in this case are as follows:

$$\hat{\alpha} = 1.1829,$$

$$\widehat{\lambda_L - \lambda_K} = 0.0173,$$

$$\hat{\sigma} = 0.1551.$$

In this chapter, as in the previous one, the correlation coefficients are accompanied in most cases by high standard errors. However, as was indicated above, alternative estimates based on cross-section data for the American case show similar estimates for σ with significantly low standard errors (see section G in the Appendix). This, to a certain extent, lessens the degree of severity of this case.

From (IIIa) and (IIIb) both estimates for $(\lambda_L - \lambda_K)$ are positive and both estimates for σ are less than unity. These clearly imply that in the wool industry in the U.S. during 1850-1900 there was a bias of labor-saving technological growth. The labor-augmentation rate of growth exceeded that of capital by approximately one half of a percentage point annually.

In the case of wool in the U.K., however, the result is different from what one might have expected. From (IV) it follows that the estimates for $(\lambda_L - \lambda_K)$ is positive but that for σ is less than unity. This implies that during this period, there was a labor-saving bias as in the case of the U.S. The annual rate of growth of this bias, however, was higher than in the U.S. — approximately 1.73. If this result does indeed reflect the "true" historical fact, it will simply imply that Habakkuk's implications cannot be held as a general rule for the combined manufacturing industry but rather for only parts of it.

A Digression: Estimation Without Lags

In this case, as in the previous case of cotton, one may wish to estimate the bias without lags. As in the previous case, the resulting estimates by fitting equation (13) to our observations vary very slightly from those obtained by the use of equation (16).

The results of fitting equation (13) to observations on k_1, k_2 and $\pi_L/1-\pi_L$ for the American wool industry are shown by (III'_a) and (III'_b) as follows:

$$(\mathrm{III}'_a)\quad \ln(k_1)_t = \underset{}{2.7619} + \underset{(0.3696)}{0.3197} \ln\left(\frac{\pi_L}{1-\pi_L}\right)_t + \underset{(0.0037)}{0.0092t};$$

and with

$$R = 0.9130,$$

$$\text{Std. error est.} = 0.0952,$$

$$\text{Degrees of freedom} = 3.$$

Thus we obtain:

$$\hat{\sigma} = 0.2423,$$

$$\lambda_L \hat{-} \lambda_K = 0.0092.$$

$$(\mathrm{III}'_b)\quad \ln(k_2)_t = 0.9578 + \underset{(0.1076)}{0.1735} \ln\left(\frac{\pi_L}{1-\pi_L}\right)_t + \underset{(0.0011)}{0.0056t};$$

and with

$$R = 0.9768,$$

$$\text{Std. error est.} = 0.0277,$$

$$\text{Degrees of freedom} = 3.$$

so that:

$$\hat{\sigma} = 0.1478,$$

$$\lambda_L \hat{-} \lambda_K = 0.0056.$$

The resulting fit for the British wool industry is given by (IV') as follows:

$$(\mathrm{IV}')\quad \ln(k_2)_t = -1.3934 + \underset{(0.5895)}{0.2580} \ln\left(\frac{\pi_L}{1-\pi_L}\right)_t + \underset{(0.0070)}{0.0209t};$$

and with

$$R = 0.8684,$$

$$\text{Std. error est.} = 0.2156,$$

$$\text{Degrees of freedom} = 3.$$

Thus we get:

$$\hat{\sigma} = 0.2051,$$

$$\widehat{\lambda_L - \lambda_K} = 0.0209.$$

The conclusions remain essentially the same. As in the previous section, it is clear that in both countries there was a labor-saving bias during the period 1850-1900. This time, however, the bias is slightly higher in both countries. But it is still higher in the U.K.

CHAPTER 9

Conclusion

The main question of this part was whether there exists empirical evidence to support Habakkuk's conclusions. The application of the C.E.S. production Model showed that both of the textile industries in the U.S. were experiencing labor-saving bias in technological growth. This result is in accord with what one might have expected to happen for the U.S. economy. We also concluded that for the U.K. cotton branch one may still retain Habakkuk's implication. In the case of U.K. wool, however, we found that contrary to what we expected, labor-saving rather than capital-saving bias was measured.[73]

We could have given an alternative interpretation if we had followed the alternative Hicksian definition of biased factor-saving technological change. As an alternative to the definition given above, one may define labor-saving neutral or capital-saving "inventions" according to whether — given a constant factor-price ratio — they increase capital intensity, leave it unchanged, or decrease capital intensity. In terms of our model, then, higher (lower) labor-augmenting bias in technology is synonymus with higher (lower) rate of growth of capital intensity

---

73. We would like to stress again that most of the parameter estimates (based on time series) were associated with relatively high standard errors. Thus the resulting conclusions tend to be more ambiguous.

provided that the elasticity of substitution is less than unity.[74]

Under the alternative Hicksian definitions then, our resulting estimates did not reject the implication that the U.S. was having a higher rate of growth of capital-intensity at least in the case of cotton, during the period of the second half of the nineteenth century. In the case of wool, we can say by the same analogy, that the British sector was having a higher rate of growth of capital intensity during this period. Both of these results, however, do not necessarily imply full contradiction of the results of the first part. Because one cannot say a priori that within this period of time a higher rate of growth of capital intensity in one country implies a higher actual capital-intensity there.[75]

---

74. This follows from equation (10) in section D of the Appendix. Differentiating with respect to time yields (10a) as follows:

$$\text{(10a)} \quad \frac{\dot{K}}{K} - \frac{\dot{L}}{L} = \sigma\left(\frac{\dot{w}}{w} - \frac{\dot{r}}{r}\right) + (1-\sigma)\left(\frac{\dot{E}_L}{E_L} - \frac{\dot{E}_K}{E_K}\right).$$

It can readily be seen that if $\sigma < 1$, the alternative Hicksian definition of a labor-saving change in technology

$$\frac{\dot{K}}{K} - \frac{\dot{L}}{L} = (1-\sigma)\left(\frac{\dot{E}_L}{E_L} - \frac{\dot{E}_K}{E_K}\right) > 0$$

is satisfied when $\dot{E}_L/E_L > \dot{E}_K/E_K$, and the alternative Hicksian definition of capital-saving change is satisfied when $\dot{E}_L/E_L < \dot{E}_K/E_K$. When $\dot{E}_L/E_L = \dot{E}_K/E_K$, we have Hicks-neutrality $\frac{\dot{K}}{K} = \frac{\dot{L}}{L}$ for all non-negative values of σ.

75. To illustrate this point we may use the case of the wool industry. We have concluded that the British sector had a higher labor-saving bias. Alternatively, one can say that the British sector had a higher rate of growth of capital-intensity. This, however, does not imply that actual capital-intensity at a given period of time was higher in the British sector. Thus, the conclusion reached in the first part of this study that actual capital-intensity in the American wool industry was higher during the period of study, does not necessarily contradict the former conclusion that labor-saving bias was lower there.

As by-products of this study, we obtained estimates for the elasticities of substitution for each sector. Many have attempted estimations of elasticities of substitution for the U.S. economy and its subsectors.[76] Our resulting estimates did not confirm the specification of constant unitary elasticity of substitution, namely, Cobb-Douglas form, for aggregate production function for either of the sectors in question. However, in the American cases, these estimates are quite low and may justify the use of production functions with zero elasticity. More specifically, our estimates show elasticity in the neighborhood of 0.05-0.1 for the U.S. cotton and 0.20-0.38 for the U.S. wool industries. In the U.K. we obtained estimates of 0.65-2.27 and 0.16 for the cotton and wool respectively.

Finally, we also obtained estimates for the rates of growth of factor-efficiency, λ's. The estimates for the rates of λ_L are given in the Appendix (section F). The calculation of $\hat{\lambda}_K$ is straightforward, namely, it is given by $\hat{\lambda} - (\lambda_L - \hat{\lambda}_K)$, which counts for its omission from the Appendix.

---

76. See summary and evaluation of these estimates in Nerlove, *ibid*.

APPENDIX

A. Short-Term Interest-Rates in the U. S. and the U. K. During the Nineteenth Century

Table A1:

Period	U.S. Call Money 10 yr. av. % (1)	U.K. Open Market Rate of Discount % (2)	Ratio (1)/(2)
1850-59	6.30	3.84	1.64
1860-69	6.50	4.18	1.55
1870-79	5.73	2.96	1.93
1880-89	3.98	2.57	1.54
1890-91	3.32	2.09	1.58

The sources are as follows:

column (1) - S. Homer, *ibid.* pp. 319-20

column (2) - S. Homer, *ibid.* pp. 208-90.

The series on call money rates are from the New York Stock Exchange in terms of monthly averages. Call money rates are for short-term loans which can be claimed by demand and paid within 24-48 hours.

The series for the open market rate of discount are annual averages for first class commercial bills with 3-months period of maturity. Because of strong fluctuations in the annual averages, we have calculated their 10-year-averages. Smoothing their extreme deviations would make their comparisons more meaningful.

B. Nominal and Real Wage-Rates in the U.S. and the U.K. During the Nineteenth Century

In the theoretical model the concept of "real-wage" is identical with marginal physical productivity of labor. Thus, the series of real weekly earnings in chapter 2 correspond to the latter notion. However, real-wages are measured in terms of a "basket" of goods and services, namely, by deflating money-wages by the cost of living index. One may wish to use, then, this concept in the context of our argument. In the following sections we shall present a series of this type. Here money-wages are deflated by the cost of living indexes (Table A2). The real-wage series for the cotton industry are given in Tables A3-A7; those for the wool are in Tables A8-A12; and those for iron and steel are in Tables A13-A16.

TABLE A2: Cost of Living Indexes in the U. S. and the U. K.

During 1846-1900

Year	Cost of Living Index U.S.(1)	U.K.(2)	Year	Cost of Living Index U.S.(1)	U.K.(2)
1846	58	106	1867	102	114
1847	58	110	1868	98	113
1848	54	96	1869	95	111
1849	51	93	1870	91	110
1850	54	94	1871	89	113
1851	60	94	1872	90	120
1852	60	96	1873	88	122
1853	64	108	1874	88	115
1854	64	115	1875	86	111
1855	67	114	1876	81	110
1856	68	114	1877	80	110
1857	70	117	1878	80	104
1858	69	106	1879	79	101
1859	63	108	1880	80	105
1860	61	113	1881	83	103
1861	63	112	1882	86	102
1862	69	113	1883	81	102
1863	79	115	1884	77	97
1864	95	115	1885	75	91
1865	102	113	1886	76	89
1866	103	114	1887	77	88

TABLE A2: (concluded)

Year	Cost of Living Index U.S.(1)	U.K.(2)	Year	Cost of Living Index U.S.(1)	U.K.(2)
1888	78	88	1895	73	83
1889	78	89	1896	74	83
1890	78	89	1897	75	85
1891	76	89	1898	75	88
1892	77	90	1899	77	86
1893	75	89	1900	80	91
1894	73	85	1913	100	100

Source: Column (1): *Historical Statistics of the U.S. 1789-1945. A Supplement to the Statistical Abstracts of the U.S.* Prepared by the Bureau of the Census with the cooperation of the Social Science Research Council, U.S. Department of Commerce 1949. These series were calculated by the Federal Reserve in New York and go back to 1820.

Column (2): A.L. Bowley, *Wages and Income in the U.K. Since 1860*. Cambridge, University Press, 1937, pp. 121-122.

TABLE A3: Nominal and Real Weekly Earnings in the Cotton Industry in the U.S. and the U.K. During the Nineteenth Century (1848-1893): Overall Averages

U.S. $

Year	Nominal Earnings U.S. (1)	Nominal Earnings U.K. (2)	Real Earnings U.S. (3)	Real Earnings U.K. (4)	Ratio (3)/(4)
1848	3.76	2.51	6.96	2.61	2.67
1849	3.68	2.62	7.20	2.82	2.55
1850	3.81	2.51	7.05	2.67	2.64
1853	3.75	2.71	5.85	2.51	2.33
1854	3.75	2.50	5.85	2.17	2.70
1859	3.64	2.83	5.77	2.62	2.20
1860	3.71	2.80	6.07	2.48	2.45
1865	5.20	2.60	5.10	2.30	2.22
1870	6.01	3.11	6.60	2.83	2.33
1871	6.07	3.16	6.82	2.80	2.44
1873	6.47	3.27	7.35	2.68	2.74
1874	5.96	3.42	6.77	2.97	2.28
1876	5.53	3.34	6.82	3.04	2.24
1877	5.13	3.87	6.40	3.52	1.82
1880	4.89	3.76	6.10	3.58	1.70
1882	5.47	3.92	6.35	3.84	1.65
1883	5.53	3.86	6.82	3.78	1.80
1884	5.38	3.93	6.98	4.05	1.72
1886	5.31	3.67	6.98	4.12	1.69
1893	6.30	3.93	8.40	4.42	1.90

<u>Source</u>:

Columns (1)-(2) See Note (33).

Notes: (a) The prevailing formal exchange-rate is used to calculate British wages in terms of U.S. dollars. Indeed, there were slight deviations from the formal rate, but they would not have affected the results. The interested reader can find estimates of the actual deviations for most of the nineteenth century in L. E. Davis and J. R. Hughes, "A Dollar-Sterling Exchange, 1803-1895", <u>The Econ. Hist. Rev</u>. Vol. XIII 1960-61.

(b) Slight differences would have occurred if hourly earnings had been taken; but even doing so, the argument would have remained valid and moreover, strengthened, because the American worker has on the average a shorter work-day for parallel periods.

TABLE A4: Nominal and Real Weekly Earnings in the Cotton Industry in the U.S. and the U.K. During the Nineteenth Century (1850-1886): Piecers

U.S. $

Year	Nominal Earnings U.S. (1)	Nominal Earnings U.K. (2)	Real Earnings U.S. (3)	Real Earnings U.K. (4)	Ratio (3)/(4)
1850	2.01	1.56	3.72	1.66	2.24
1859	2.64	1.44	4.19	1.33	3.15
1860	2.64	1.56	4.33	1.38	3.14
1870	4.07	1.92	4.47	1.75	2.55
1874	4.38	1.92	4.98	1.67	2.98
1877	3.96	2.04	4.95	1.85	2.68
1880	3.18	2.04	3.98	1.94	2.05
1882	3.65	2.64	4.24	2.59	1.64
1884	3.75	1.92	4.87	1.98	2.46
1886	3.60	2.34	4.74	2.63	1.80

Source: See Note (34) and Notes (a)-(b) for Table A3 in this Appendix.

TABLE A5: Nominal and Real Weekly Earnings in the Cotton Industry in the U.S. and the U.K. During the Nineteenth Century (1850-1886): Grinders

U.S. $

Year	Nominal Earnings U.S. (1)	U.K. (2)	Real Earnings U.S. (3)	U.K. (4)	Ratio (3)/(4)
1850	4.89	3.36	9.06	3.54	2.56
1853	5.64	3.68	8.81	3.41	2.58
1854	5.64	3.36	8.81	2.92	3.02
1859	5.88	3.60	9.33	3.33	2.80
1860	6.15	3.66	10.08	3.24	3.11
1865	8.61	3.84	8.44	3.40	2.48
1870	10.92	4.34	12.00	3.95	3.04
1874	9.24	5.64	10.50	4.90	2.14
1877	7.65	5.30	9.56	4.82	1.98
1880	8.34	5.40	10.43	5.14	2.03
1882	7.68	5.04	8.93	4.94	1.81
1883	8.31	5.06	10.26	4.96	2.03
1884	8.43	4.66	10.95	4.80	2.28
1886	9.00	5.14	11.84	5.78	2.05

Source: See Note (35) and Notes (a)-(b) for Table A3 in this Appendix.

TABLE A6: Nominal and Real Weekly Earnings in the Cotton Industry in the U.S. and the U.K. During the Nineteenth Century (1849-1886): Overseers, Spinning Department

U.S. $

Year	Nominal Earnings U.S. (1)	Nominal Earnings U.K. (2)	Real Earnings U.S. (3)	Real Earnings U.K. (4)	Ratio (3)/(4)
1849	12.00	5.28	23.53	5.68	4.14
1850	11.25	5.28	20.83	5.62	3.71
1859	12.00	6.24	19.05	5.78	3.30
1860	12.00	6.24	19.67	5.52	3.56
1870	10.50	7.20	11.54	6.54	1.77
1873	15.72	7.20	17.86	5.90	3.03
1877	18.00	9.60	22.50	8.73	2.58
1880	25.50	9.60	31.88	9.14	3.49
1882	22.50	10.20	26.16	10.00	2.61
1883	24.00	12.00	29.63	11.76	2.52
1886	24.60	10.20	32.37	11.46	2.82

Source: See Note (36) and Notes (a)-(b) for Table A3 in this Appendix.

TABLE A7: Nominal and Real Weekly Earnings in the Cotton Industry in the U.S. and the U.K. During the Nineteenth Century (1849-1886): Overseers, Carding Department

U.S. $

Year	Nominal Earnings U.S. (1)	U.K. (2)	Real Earnings U.S. (3)	U.K. (4)	Ratio (3)/(4)
1849	12.00	6.72	23.53	7.23	3.25
1850	12.00	6.48	22.22	6.89	3.22
1859	15.00	6.72	23.81	6.22	3.83
1860	15.00	6.72	24.60	5.95	4.13
1870	18.00	7.68	19.80	6.99	2.83
1873	18.00	7.68	20.45	6.30	3.25
1874	18.00	9.00	20.45	7.83	2.61
1877	18.00	9.48	22.50	8.62	2.61
1880	24.00	9.24	30.00	8.80	3.41
1882	30.00	10.56	34.88	10.35	3.37
1883	30.00	9.84	37.04	9.65	3.84
1886	30.00	9.36	39.47	10.52	3.75

Source: See Note (37) and Notes (a)-(b) for Table A3 in this Appendix.

TABLE A8: Nominal and Real Weekly Earnings in the Woolen Industry in the U.S. and the U.K. During the Nineteenth Century (1850-1900): Overall Averages

U.S. $

Year	Nominal Earnings U.S. (1)	Nominal Earnings U.K. (2)	Real Earnings U.S. (3)	Real Earnings U.K. (4)	Ratio (3)/(4)
1850	4.11	2.60	7.61	2.77	2.75
1860	4.64	3.18	7.61	2.81	2.71
1870	6.84	3.30	7.52	3.00	2.51
1880	6.13	4.12	7.66	3.92	1.95
1890	7.05	3.62	9.04	4.07	2.22
1900	7.28	3.96	9.10	4.35	2.09

Source: Column (1) Census Reports. Twelfth Census of the U.S. Taken in the Year 1900. Manufactures: Textiles. Washington U.S. Census Office 1902. To get average earnings per worker, total wages were divided by average number of workers (ibid. Table 29, pp. 122-123.)

Column (2) A.L. Bowley, "The Statistics of Wages in U.K. During the Last Hundred Years, Part IX", Journal of the Royal Statistical Society, 1902. These are weighted averages based on Bowley's numbers. We have taken 1886's weights to calculate each year's average, except for the years 1860 and

1880 which are from T. Ellison's *The Cotton Trade of Great Britain* (London 1886), pp. 123-24.

See also Notes (a)-(b) for Table A3 in this Appendix.

TABLE A9: Nominal and Real Weekly Earnings in the Woolen Industry in the U.S. and the U.K. During the Nineteenth Century (1850-1900): Washers

U.S. $

Year	Nominal Earnings U.S. (1)	Nominal Earnings U.K. (2)	Real Earnings U.S. (3)	Real Earnings U.K. (4)	Ratio (3)/(4)
1850	4.98	3.54	9.22	3.77	2.45
1855	5.28	3.54	7.88	3.11	2.53
1856	5.28	3.54	7.76	3.11	2.50
1857	5.40	3.54	7.71	3.03	2.54
1858	5.52	3.84	8.00	3.62	2.21
1860	5.52	3.40	9.05	3.01	3.01
1861	5.76	3.60	9.14	3.21	2.85
1863	5.88	3.68	7.44	3.20	2.33
1866	7.74	3.96	7.51	3.47	2.16
1867	7.98	4.20	7.82	3.68	2.13
1868	7.50	4.20	7.65	3.72	2.06
1870	7.50	3.96	8.24	3.60	2.30
1871	7.50	3.96	8.43	3.50	2.41
1872	7.50	4.80	8.33	4.00	2.08
1874	6.96	4.08	7.91	3.55	2.23
1876	6.78	5.28	8.37	4.80	1.74
1877	6.00	4.32	7.50	3.93	1.91

TABLE A9: (concluded)

Year	Nominal Earnings U.S. (1)	Nominal Earnings U.K. (2)	Real Earnings U.S. (3)	Real Earnings U.K. (4)	Ratio (3)/(4)
1880	6.75	4.86	8.44	4.63	1.82
1883	6.75	5.40	8.33	5.29	1.57
1886	6.99	4.90	9.20	5.51	1.67
1890	6.99	4.80	8.96	5.40	1.66
1900	6.96	4.90	8.70	5.38	1.62

Source: See Note (39) and also Notes (a)-(b) for Table A3 in this Appendix.

TABLE A10: Nominal and Real Weekly Earnings in the Woolen Industry in the U.S. and the U.K. During the Nineteenth Century (1850-1900): Dyers

U.S. $

Year	Nominal Earnings U.S. (1)	Nominal Earnings U.K. (2)	Real Earnings U.S. (3)	Real Earnings U.K. (4)	Ratio (3)/(4)
1850	6.00	3.60	11.11	3.83	2.90
1855	6.15	3.60	9.18	3.16	2.91
1856	6.39	3.60	9.40	3.16	2.97
1857	6.48	3.60	9.26	3.08	3.00
1860	6.60	4.80	10.82	4.25	2.55
1861	7.35	4.80	11.67	4.29	2.72
1863	7.35	4.80	9.30	4.17	2.23
1866	7.65	4.32	7.43	3.79	1.96
1867	8.25	4.80	8.09	4.21	1.92
1868	8.70	4.80	8.88	4.25	2.09
1870	7.98	3.96	8.77	3.60	2.44
1871	9.00	3.96	10.11	3.50	2.89
1872	9.00	4.08	10.00	3.40	2.95
1874	9.00	5.76	10.23	5.01	2.04
1876	9.00	5.28	11.11	4.80	2.31
1877	6.72	5.28	8.40	4.80	1.75
1880	7.68	5.28	9.60	5.03	1.91

TABLE A10: (concluded)

Year	Nominal Earnings U.S. (1)	U.K. (2)	Real Earnings U.S. (3)	U.K. (4)	Ratio (3)/(4)
1886	9.00	4.66	11.84	5.24	2.26
1890	9.00	4.66	11.54	5.24	2.20
1900	9.00	5.28	11.25	5.80	1.94

Source: See Note (40) and also Notes (a)-(b) for Table A3 in this Appendix.

TABLE A11: Nominal and Real Weekly Earnings in the Woolen Industry in the U.S. and the U.K. During the Nineteenth Century (1850-1900): Sorters

U.S. $

Year	Nominal Earnings U.S. (1)	Nominal Earnings U.K. (2)	Real Earnings U.S. (3)	Real Earnings U.K. (4)	Ratio (3)/(4)
1850	6.75	2.82	12.50	6.20	2.02
1860	9.00	4.98	14.75	4.41	3.34
1861	9.00	4.98	14.29	4.45	3.21
1863	11.82	5.36	14.96	4.66	3.21
1866	14.70	6.72	14.27	5.89	2.42
1867	16.50	6.72	16.18	5.89	2.75
1868	12.63	6.72	12.89	5.95	2.17
1870	13.50	6.00	14.84	5.46	2.72
1871	13.02	6.00	14.63	5.31	2.76
1872	13.11	6.72	14.57	5.60	2.60
1874	12.93	5.48	14.69	4.77	3.08
1876	11.85	6.48	14.63	5.89	2.48
1877	12.15	7.68	15.19	6.98	2.18
1880	12.69	7.50	15.86	7.14	2.22
1883	15.21	7.40	18.78	7.25	2.59
1886	16.50	6.80	21.71	7.64	2.84
1890	13.23	7.20	16.96	8.09	2.10
1891	12.24	7.20	16.11	8.09	1.99
1900	12.27	7.20	15.34	7.91	1.94

TABLE A11:

<u>Source</u>: See Note (41) and also Notes (a)-(b) for Table A3 in this Appendix.

TABLE A12: Nominal and Real Weekly Earnings in the Woolen Industry in the U.S. and the U.K. During the Nineteenth Century (1850-1900): Overseers, Weaving Department

U.S. $

Year	Nominal Earnings U.S. (1)	Nominal Earnings U.K. (2)	Real Earnings U.S. (3)	Real Earnings U.K. (4)	Ratio (3)/(4)
1850	7.50	5.22	13.89	5.55	2.50
1860	10.50	6.32	17.21	5.59	3.08
1861	10.89	6.32	17.29	5.64	3.07
1863	15.00	6.64	18.99	5.77	3.29
1866	24.00	8.40	23.30	7.37	3.16
1867	21.00	7.20	20.59	6.36	3.24
1868	18.00	7.20	18.37	6.37	2.88
1870	21.00	6.96	23.08	6.33	3.64
1871	18.00	6.96	20.22	6.16	3.28
1874	21.00	8.16	23.86	7.10	3.36
1876	24.00	7.92	29.63	7.20	4.11
1877	18.00	8.04	22.50	7.31	3.08
1880	20.25	8.10	25.31	7.71	3.28
1883	21.00	8.16	25.93	8.00	3.24
1886	18.00	7.52	23.68	8.45	2.80
1890	18.00	7.52	23.08	8.45	2.73
1900	18.00	8.24	22.50	9.06	2.48

Source: See Note (42) and also Notes (a)-(b) for Table A3 in this Appendix.

TABLE A13: Nominal and Real Weekly Earnings in the Iron and Steel Industry in the U.S. and the U.K. During the Nineteenth Century (1840-1886):

Machinists

U.S. $

Year	Nominal Earnings U.S. (1)	U.K. (2)	Real Earnings U.S. (3)	U.K. (4)	Ratio (3)/(4)
1840	8.28	7.12	13.80	5.93	2.33
1845	8.97	7.10	16.61	6.89	2.41
1846	9.51	7.10	16.40	6.70	2.45
1849	11.40	5.72	22.35	6.15	3.63
1871	19.20	8.16	21.57	7.22	2.99
1874	19.50	8.16	22.16	7.10	3.12
1884	16.68	7.68	21.66	7.92	2.73
1886	16.68	7.88	21.95	8.85	2.48

Source: See Note (44) and also Notes (a)-(b) for Table A3 in this Appendix.

TABLE A14: Nominal and Real Weekly Earnings in the Iron and Steel Industry in the U.S. and the U.K. During the Nineteenth Century (1859-1886):

Millwrights

U.S. $

Year	Nominal Earnings U.S. (1)	Nominal Earnings U.K. (2)	Real Earnings U.S. (3)	Real Earnings U.K. (4)	Ratio (3)/(4)
1859	12.00	7.68	19.05	7.11	2.68
1871	21.00	7.20	23.60	6.37	3.70
1877	16.50	7.68	20.63	6.98	2.96
1880	15.00	7.44	18.75	7.09	2.64
1884	16.80	8.16	21.82	8.41	2.59
1886	15.00	8.04	19.74	9.03	2.19

Source: See Note (45) and also Notes (a)-(b) for Table A3 in this Appendix.

TABLE A15: Nominal and Real Weekly Earnings in the Iron and Steel Industry in the U.S. and the U.K. During the Nineteenth Century (1859-1886):

Boilermakers

U.S. $

Year	Nominal Earnings U.S. (1)	U.K. (2)	Real Earnings U.S. (3)	U.K. (4)	Ratio (3)/(4)
1859	12.48	7.68	19.81	7.11	2.79
1860	11.64	7.20	19.08	6.37	3.00
1864	13.38	8.04	14.08	6.99	2.01
1870	14.22	7.20	15.63	6.55	2.39
1874	18.99	8.46	21.58	7.36	2.93
1877	14.04	8.16	17.55	7.42	2.37
1883	17.49	7.68	21.59	7.53	2.87
1884	17.49	8.08	22.71	8.33	2.73
1886	17.49	7.72	23.01	8.67	2.65

Source: See Note (46) and also Notes (a)-(b) for Table A3 in this Appendix.

TABLE A16: Nominal and Real Weekly Earnings in the Iron and Steel Industry in the U.S. and the U.K. During the Nineteenth Century (1849-1886):

Patternmakers

U.S. $

Year	Nominal Earnings U.S. (1)	Nominal Earnings U.K. (2)	Real Earnings U.S. (3)	Real Earnings U.K. (4)	Ratio (3)/(4)
1849	9.63	7.20	18.88	7.74	2.44
1851	10.38	7.32	17.30	7.79	2.22
1859	11.10	7.68	17.61	7.11	2.48
1864	15.93	7.56	16.77	6.57	2.55
1871	20.46	8.88	22.99	7.86	2.92
1877	15.78	7.44	19.73	6.76	2.92
1880	15.90	7.20	19.88	6.86	2.90
1883	20.34	8.40	25.11	8.24	3.05
1884	21.00	8.88	27.27	9.15	2.98
1886	19.62	8.32	25.82	9.35	2.76

Source: See Note (47) and also Notes (a)-(b) for Table A3 in this Appendix.

C. "Real-Wages" in the Iron and Steel Industry in the U.S. and the U.K. During the Nineteenth Century

In section D of chapter 2 (Part I) MPP of labor (or "real-wages") were given in terms of bar iron per-worker-per-week. Here the price deflator is pig iron, so the MPP of labor is given in terms of pig iron per-worker-per-week. These series are given in Tables A17-A20.

TABLE A17: Average Weekly Earnings in the Iron and Steel Industry in the U.S. and the U.K. During the Nineteenth Century (1846-1886):

Machinists

Year	Nominal Earnings U.S. $(1)	Nominal Earnings U.K. s/ d(2)	Av. Price of Pig Iron per Ton U.S. $(3)	Av. Price of Pig Iron per Ton U.K. L/ s/d(4)	Iron per Worker per Week (tons) U.S. (5)	Iron per Worker per Week (tons) U.K. (6)	Ratio (5)/(6)
1846	9.51	29/ 7	27.87	3/ 7/ 3	0.341	0.440	0.78
1849	11.40	23/10	22.75	2/ 5/ 6	0.501	0.523	0.96
1871	19.20	34/ 0	35.12	2/18/11	0.547	0.577	0.95
1874	19.50	34/ 0	30.25	4/ 7/ 6	0.645	0.389	1.66
1884	16.68	32/ 0	19.87	2/ 2/ 1	0.839	0.760	1.10
1886	16.86	32/10	18.71	1/19/11	0.901	0.802	1.12

Source: Columns (1)-(2): See Note (44).

Column (3): _Historical Statistics of the U.S. Colonial Times to 1957._ U.S. Bureau of the Census, Washington, 1961, pp. 365-366.

Column (4): Mitchell and Deane, _op. cit._, pp. 492-93.

TABLE A18: Average Weekly Earnings in the Iron and Steel Industry in the U.S. and the U.K. During the Nineteenth Century (1859-1886):

Millwrights

Year	Nominal Earnings U.S. $(1)	U.K. s/d(2)	Av. Price of Pig Iron per Ton U.S. $(3)	U.K. L/ s/d(4)	Iron per Worker per Week (tons) U.S. (5)	U.K. (6)	Ratio (5)/(6)
1859	12.00	32/0	23.37	2/11/ 9	0.513	0.618	0.83
1871	21.00	30/0	35.12	2/18/11	0.598	0.509	1.17
1877	16.50	32/0	18.87	2/14/ 4	0.874	0.588	1.48
1880	15.00	31/0	28.50	2/14/ 6	0.526	0.568	0.92
1884	16.80	34/0	19.87	2/ 2/ 1	0.845	0.771	1.09
1886	15.00	33/6	18.71	1/19/11	0.825	0.839	0.98

Source: Columns (1)-(2): See Note (45).

Columns (3)-(4): See source for Table A17 in this Appendix.

TABLE A19: Average Weekly Earnings in the Iron and Steel Industry in the U.S. and the U.K. During the Nineteenth Century (1859-1886):

Boilermakers

Year	Nominal Earnings U.S. $(1)	Nominal Earnings U.K. s/d(2)	Av. Price of Pig Iron per Ton U.S. $(3)	Av. Price of Pig Iron per Ton U.K. L/ s/d(4)	Iron per Worker per Week (tons) U.S. (5)	Iron per Worker per Week (tons) U.K. (6)	Ratio (5)/(6)
1859	12.48	32/0	23.37	2/11/ 9	0.534	0.618	0.86
1860	11.64	30/0	22.75	2/13/ 6	0.512	0.560	0.91
1864	13.88	33/6	59.25	2/17/ 3	0.225	0.585	0.38
1870	14.22	30/0	33.25	2/14/ 4	0.428	0.552	0.77
1874	18.99	35/8	30.25	4/ 7/ 6	0.628	0.407	1.54
1877	14.04	34/0	18.87	2/14/ 4	0.744	0.625	1.19
1883	17.49	32/0	22.37	2/ 6/ 9	0.782	0.684	1.14
1884	17.49	33/8	19.87	2/ 2/ 1	0.880	0.801	1.09
1886	17.49	32/2	18.71	1/19/11	0.934	0.805	1.16

Source: Columns (1)-(2): See Note (46).

Columns (3)-(4): See source for Table A17 in this Appendix.

TABLE A20: Average Weekly Earnings in the Iron and Steel Industry in the U.S. and the U.K. During the Nineteenth Century (1849-1886):

Patternmakers

Year	Nominal Earnings U.S. $(1)	U.K. s/d(2)	Av. Price of Pig Iron per Ton U.S. $(3)	U.K. L/ s/d(4)	Iron per Worker per Week (tons) U.S. (5)	U.K. (6)	Ratio (5)/(6)
1849	9.63	30/0	22.75	2/ 5/ 6	0.423	0.659	0.64
1851	10.38	30/6	21.37	1/19/ 9	0.485	0.767	0.63
1859	11.10	32/0	23.37	2/11/ 9	0.475	0.618	0.76
1864	15.93	31/6	59.25	2/17/ 3	0.269	0.550	0.48
1871	20.46	37/0	35.12	2/18/11	0.583	0.628	0.92
1877	15.78	31/0	18.87	2/14/ 4	0.836	0.570	1.44
1880	15.90	30/0	28.50	2/14/ 6	0.558	0.550	1.01
1883	20.34	35/0	22.37	2/ 6/ 9	0.909	0.748	1.21
1884	21.00	37/0	19.87	2/ 2/ 1	1.057	0.839	1.26
1886	19.62	34/8	18.71	1/19/11	1.048	0.868	1.20

Source: Columns (1)-(2): See Note (47).

Columns (3)-(4): See source for Table A17 in this Appendix.

D. An Aggregate Production Function with Labor- and Capital-Augmenting Technological Change

Let

$$(1) \quad V = [(E_L L)^{-\rho} + (E_K K)^{-\rho}]^{-1/\rho} \; ,$$

be our production function. (1) is homogeneous of the first degree in inputs of capital and labor measured in efficiency units. This function is characterized by constant elasticity of substitution between the inputs, namely,

$$0 \leq \sigma = \frac{1}{1+\rho} = \text{constant}.$$

Alternatively, the relations between the elasticity of substitution, σ, and ρ is as follows:

$$-1 \leq \rho = \frac{1-\sigma}{\sigma} = \text{constant}.$$

We also assume that the first and second partial derivatives of the output, V, with respect to the inputs of labor and capital services ($E_L L$ and $E_K K$ respectively) are continuous positive and continuous negative respectively. The coefficients E_L and E_K represent the levels of efficiency of labor and capital respectively. L and K, then, represent conventional measures of the physical flow of labor and capital inputs, namely, labor is measured as man-years employed and capital is measured in terms of machines of a constant kind.

Now by differentiating (1) partially with respect to L we get

$$m_L \equiv \frac{\partial V}{\partial L} = -\frac{1}{\rho}Q^{-\frac{1}{\rho}-1} - \rho E_L{}^{-\rho} L^{-\rho-1} ,$$

$$= E_L{}^{-\rho} L^{-(1+\rho)} \frac{V}{Q} ,$$

$$= E_L{}^{-\rho} L^{-(1+\rho)} \frac{V}{V^{-\rho}}$$

$$= E_L{}^{-\rho} L^{-(1+\rho)} V^{1+\rho} ,$$

(2) $\therefore m_L = E_L{}^{-\rho} (\frac{V}{L})^{1+\rho}$;

where $Q \equiv [(E_L L)^{-\rho} + (E_K K)^{-\rho}]$.

Similarly we get

(3) $m_K \equiv \frac{\partial V}{\partial K} = E_K{}^{-\rho} (\frac{V}{K})^{1+\rho}$

Now differentiating (1), (2) and (3) with respect to time we get (4), (5) and (6) as follows:

(4) $\frac{\dot{V}}{V} = \alpha(\frac{\dot{L}}{L} + \frac{\dot{E}_L}{E_L}) + \beta(\frac{\dot{K}}{K} + \frac{\dot{E}_K}{E_K})$;

where α and β represent the elasticities of output with respect to L and K respectively.

(5) $\frac{\dot{m}_L}{m_L} = \frac{\sigma-1}{\sigma} \frac{\dot{E}_L}{E_L} + \frac{1}{\sigma} (\frac{\dot{V}}{V} - \frac{\dot{L}}{L})$;

(6) $\frac{\dot{m}_K}{m_K} = \frac{\sigma-1}{\sigma} \frac{\dot{E}_K}{E_K} + \frac{1}{\sigma} (\frac{\dot{V}}{V} - \frac{\dot{K}}{K})$.

Now by substituting from (4) into (5) and (6) we get (7) and (8) as follows:

(7) $\frac{\dot{m}_L}{m_L} = \frac{\beta}{\sigma} (\frac{\dot{K}}{K} - \frac{\dot{L}}{L}) + \frac{1}{\sigma} \frac{\dot{E}}{E} + \frac{\sigma-1}{\sigma} \frac{\dot{E}_L}{E_L}$;

$$(8) \qquad \frac{\dot{m}_K}{m_K} = -\frac{\alpha}{\sigma}\left(\frac{\dot{K}}{K} - \frac{\dot{L}}{L}\right) + \frac{1}{\sigma}\frac{E}{E} + \frac{\sigma-1}{\sigma}\frac{\dot{E}_K}{E_K} \; ;$$

where

$$\frac{\dot{E}}{E} \equiv \alpha\frac{\dot{E}_L}{E_L} + \beta\frac{\dot{E}_K}{E_K}$$

Now subtracting (8) from (7) we get

$$(9) \qquad \frac{\dot{m}_L}{m_L} - \frac{\dot{m}_K}{m_K} = \left(\frac{\dot{K}}{K} - \frac{\dot{L}}{L}\right)\left(\frac{\beta+\alpha}{\sigma}\right) + \frac{\sigma-1}{\sigma}\left(\frac{\dot{E}_L}{E_L} - \frac{\dot{E}_K}{E_K}\right).$$

It is clear that for a given constant capital-labor ratio the expression (9) becomes

$$(9a) \qquad \frac{\dot{m}_L}{m_L} - \frac{\dot{m}_K}{m_K} = \frac{\sigma-1}{\sigma}\left(\frac{\dot{E}_L}{E_L} - \frac{\dot{E}_K}{E_K}\right).$$

Equation (9a) states relationships between rates of change of marginal physical productivities and rates of change in levels of factor-efficiency at a given constant capital-labor ratio. Thus it can be used to relate the Hicksian definitions of factor-augmenting technological progress and the size of σ.

Now assuming perfect competition and profit maximization, one can take the real wage, w, as equal to m_L; and the real rate of return on capital, r, as equal to m_K. Also for α and β one can substitute the share of labor, π_L, and the share of capital, π_K, respectively.

Now substituting w and r for m_L and m_K respectively and dividing (2) by (3) we obtain:

$$\frac{w}{r} = (\frac{E_L}{E_K})^{-\rho}(\frac{V}{L}/\frac{V}{K})^{1+\rho} ,$$

$$= (\frac{E_L}{E_K})^{-\rho}(\frac{K}{L})^{1+\rho} ,$$

$$= (\frac{E_L}{E_K})^{\frac{\sigma-1}{\sigma}} (\frac{K}{L})(\frac{K}{L})^{\frac{1-\sigma}{\sigma}}$$

$$\therefore (\frac{w}{r})^{\sigma} = (\frac{E_L}{E_K})^{\sigma-1} (\frac{K}{L}) ;$$

so that

$$(10) \qquad \frac{K}{L} = (\frac{w}{r})^{\sigma} (\frac{E_L}{E_K})^{1-\sigma} .$$

By multiplying both sides of (10) by $(L/K)^{\sigma}$ we get:

$$(\frac{L}{K})^{\sigma}(\frac{K}{L}) = (\frac{L}{K})^{\sigma}(\frac{w}{r})^{\sigma}(\frac{E_L}{E_K})^{1-\sigma},$$

$$\therefore (\frac{K}{L})^{1-\sigma} = (\frac{Lw}{Kr})^{\sigma}(\frac{E_L}{E_K})^{1-\sigma} ;$$

so that

$$(11) \qquad \frac{K}{L} = (\frac{\pi_L}{\pi_K})^{\frac{\sigma}{1-\sigma}} \frac{E_L}{E_K} .$$

Now if we assume that changes in the relative efficiency level of labor grow over time at a constant geometrical rate $(\lambda_L-\lambda_K)$ given by

$$(12) \qquad \left(\frac{E_L}{E_K}\right)_t = \frac{E_L(0)}{E_K(0)} e^{(\lambda_L-\lambda_K)t} \quad ;$$

and substituting into (11) will give us the following natural logarithmic expression:

$$(13) \qquad \ln\left(\frac{K}{L}\right) = \frac{\sigma}{1-\sigma} \ln\left(\frac{\pi_L}{\pi_K}\right) + (\lambda_L-\lambda_K)t + \ln\left(\frac{E_L(0)}{E_K(0)}\right)$$

Equation (13) provides us with the basic least-squares regression model for our study in estimating both the elasticity of substitution and the magnitude of the bias in efficiency growth.

This regression model can estimate only the magnitude of the bias but not the level of factor-efficiency rate of growth. One, however, can develop a regression model which will allow us such estimations. From (2) we readily get by substitution (2a) as follows:

$$(2a) \qquad w^{\sigma} = E_L^{\sigma-1} \frac{V}{L} \, ,$$

$$\therefore \qquad \frac{w^{\sigma}}{w} = E_L^{\sigma-1} \frac{V}{Lw} \quad ;$$

so that

$$(14) \qquad \pi_L = E_L^{\sigma-1} w^{1-\sigma}$$

If we again specify that the efficiency of labor, E_L, grows through time at the exponential rate λ_L, equation (14) leads us to the following natural logarithmic expression:

$$(15) \qquad \ln \pi_L = (1-\sigma) \ln w + \lambda_L(\sigma-1)t + \ln E_L(0).$$

From (15) one can estimate the parameters of λ_L and σ. Similarly from (3) one can develop a regression model to estimate λ_K. However, knowing λ_L enables us to calculate λ_K simply as a residual, namely, $\hat{\lambda}_K = \hat{\lambda}_L - (\widehat{\lambda_L - \lambda_K})$ [because of the previous estimation of $(\lambda_L-\lambda_K)$ from the regression model given by (13)].

E. Factor-Shares in the U.S. and the U.K. During the Nineteenth Century

Factor-shares are calculated for each sector as fractions of value added. Labor-share is a ratio between the annual wage-bill and the corresponding value-added of the industry. Capital-share is defined, on the other hand, as unity minus labor-share. For each sector, value-added is calculated as the difference between the annual value of output and the cost of the main raw-materials (i.e., cotton and wool respectively) for the same year. The resulting series of factor-shares are given in the following Tables (capital-shares are to be obtained, of course, by subtracting the ratios in column (5) from unity).

TABLE A21: Factor-Shares in the Cotton Industry in the U.S. During the Nineteenth Century (1850-1900)

$ 1000's

Year	Value of Output (1)	Cost of Raw Cotton (2)	"Value-Added" (3)	Total Wages (4)	Labor-Share (4)/(3) (5)
1850	61,869	31,352	30,518	17,267	0.566
1860	115,682	51,557	64,125	23,940	0.373
1870	177,490	100,563	76,926	39,044	0.508
1880	192,090	86,946	105,144	42,041	0.400
1890	267,982	117,393	150,589	66,025	0.438
1900	339,200	125,170	214,031	86,690	0.405

Source: Census Reports: Textiles. ibid., pp. 54-55. Until 1870, the cost of cotton was estimated as approximately 90% of total cost of material used.

TABLE A22: Factor-Shares in the Cotton Industry in the U.K.
During the Nineteenth Century (1820-1880)

L 1000's

Year	Value of Output (1)	Cost of Raw Cotton (2)	"Value-Added" (3)	Total Wages (4)	Labor-Share (4)/(3) (5)
1820	29,388	6,187	23,201	10,413	0.449
1830	32,062	6,713	25,349	11,604	0.458
1840	41,797	10,200	31,598	13,618	0.431
1850	57,854	17,725	40,128	16,651	0.415
1860	80,231	29,290	50,941	20,704	0.406
1870	93,154	33,750	59,404	24,873	0.418
1880	106,077	38,211	67,866	29,042	0.428

Source: Ellison *ibid.*, p. 69; the values for 1840, 1850 and 1870 are estimates based on Ellison's numbers.

TABLE A23: Factor-Shares in the Wool Industry in the U.S. During the Nineteenth Century (1850-1900)

$ 1000's

Year	Value of Output (1)	Cost of Woolen Raw Materials(2)	"Value-Added" (3)	Total Wages (4)	Labor-Share (4)/(3) (5)
1850	48,609	28,832	19,777	9,334	0.472
1860	73,454	43,447	30,007	11,700	0.390
1870	199,257	124,319	74,938	35,928	0.479
1880	238,086	149,161	88,925	40,688	0.458
1890	270,528	167,234	103,294	54,340	0.526
1900	296,990	181,159	115,831	57,934	0.500

Source: Census Reports: Textiles. ibid., pp. 122-23. Column (2) accounts for slight expenses on cotton raw material

TABLE A24: Factor-Shares in the Wool Industry in the U.K. During the Nineteenth Century (1850-1900)

L 1000's

Year	Value of Output (1)	Cost of Woolen Raw Materials(2)	"Value-Added" (3)	Total Wages (4)	Labor-Share (4)/(3) (5)
1850	36,400	13,100	23,300	6,879	0.295
1860	47,600	20,100	27,500	8,937	0.325
1870	59,600	24,500	35,100	9,040	0.258
1880	56,600	20,300	36,300	13,556	0.373
1890	59,700	19,600	40,100	11,388	0.284
1900	63,400	19,600	43,800	10,725	0.245

Source: Deane and Cole *ibid*., p. 196. The total-wages for 1860 and 1880 are from Ellison ibid. pp. 123-24; the rest are based on Bowley's article in J.R.S.S. 1902 ibid. To estimate these numbers we used Bowley's weights for the year 1886.

F. Parameter Estimates for the Cotton and Wool Industries in the U.S. and the U.K. During the Nineteenth Century

In this section we shall try to estimate the levels of λ_L and λ_K for both sectors. It will be sufficient, however, to estimate one of these parameters in order to know the estimate of the other since we already know the difference $(\lambda_L - \lambda_K)$ from chapters 7-8 (Part II).

For this purpose we shall use the model given by (15) in section D. This simply requires fitting observations on π_L and w. Since we have presented several series of wages, one may be confronted with the problem of which of the w's to choose. The resulting estimates will be different for different wage series. In this section we shall present the resulting estimates of those wage series which yielded the closest estimate for σ to the one previously obtained.

Fitting, then, equation (15) by least-squares regression to the data for the cotton industry in the U.S. during 1850-1900 yields the following results:

$$\text{(a)} \quad \ln(\pi_L)_t = -3.0022 + \underset{(0.4820)}{0.5576} \ln(w)_t - \underset{(0.0116)}{0.0168t} \; ;$$

and with

$$R = 0.6839,$$

$$\text{Std. error est.} = 0.1162,$$

$$\text{Degrees of freedom} = 3.$$

The resulting parameter estimates are

$$\hat{\sigma} = 0.4424,$$

$$\hat{\lambda}_L = 0.038.$$

The wage-series for the estimation of (a) are those of overall averages for the industry in terms of cloth-per-worker-per-week (see Table 2 in Part I).

Similarly fitting equation (15) by least-squares regression to data for the cotton industry in the U.K. during 1820-1880, we get the following result:

$$\text{(b)} \quad \ln(\pi)_t = -0.4273 - \underset{(0.1573)}{0.1519} \ln(w)_t + \underset{(0.0031)}{0.0015t};$$

and with

$$R = 0.6984,$$

$$\text{Std. error est.} = 0.0352,$$

$$\text{Degrees of freedom} = 4.$$

The resulting estimates are

$$\hat{\sigma} = 1.1519,$$

$$\hat{\lambda}_L = 0.0099.$$

The wage-series for this estimation are from "Piecers" in terms of cloth-per-worker-per-week (see Table 3 in Part I).

The result for the wool industry in the U.S. during 1850-1900 is given by (c) as follows:

(c) $$\ln(\pi_L)_t = -0.6069 - \underset{(0.3324)}{0.0836}\ln(w)_t + \underset{(0.0041)}{0.0041}t \; ;$$

and with

$$R = 0.6073,$$

$$\text{Std. error est.} = 0.0813,$$

$$\text{Degrees of freedom} = 3.$$

The resulting parameter estimates are

$$\hat{\sigma} = 1.0836,$$

$$\hat{\lambda}_L = 0.0490.$$

The wage-series for this estimation are from "Overseers" (see Table A12 in the Appendix).

For the wool industry in the U.K. during 1850-1900 the following result is obtained:

(d) $$\ln(\pi_L)_t = -5.6790 + \underset{(0.7176)}{0.8679}\ln(w)_t - \underset{(0.0072)}{0.0193}t \; ;$$

and with

$$R = 0.6348,$$

$$\text{Std. error est.} = 0.1194,$$

$$\text{Degrees of freedom} = 3.$$

The resulting parameter estimates are

$$\hat{\sigma} = 0.1321,$$

$$\hat{\lambda}_L = 0.0119.$$

The wage-series for this estimation are from "Dyers" (see Table A10 in the Appendix).

G. Elasticity Estimates Based on Cross-Section Data: American Cotton and Wool Industries in the Nineteenth Century

Since the standard errors accompanying the regression coefficients in chapters 7 and 8 (Part II) were significantly high, it is desirable to see whether estimates for σ resulting from cross-section data differ significantly from those already obtained. Unfortunately, data is available only for the U.S. Thus, the test is not performed for the British sectors.

The procedure involved is quite simple. All we have to do is to fit equation (11) (see section D in this Appendix) to observations on k_1, k_2 and $\pi_L/1-\pi_L$ for each of the census years. Before presenting the results we should mention that the source of data remains unchanged, namely, "Census Reports, Manufactures - Textiles" (1902, *ibid.*, pp. 54-59 for cotton, and pp. 122-29 for wool). The period covered is 1860-1900. The resulting regressions (with non-negative estimates for σ) are presented by (1)-(9) and by (10)-(14) for cotton and wool respectively.

A. The Cotton Sector

For the year 1900, the results of fitting equation (11) to observations on k_1, k_2 and $\pi_L/1-\pi_L$ are given by (1) and (2) as follows:

$$(1) \quad \ln(k_1) = 3.9688 + \underset{(0.2406)}{0.6413} \ln\left(\frac{\pi_L}{1-\pi_L}\right) ;$$

and with

$$R = 0.5120,$$

$$\text{Std. error est.} = 0.3370,$$

$$\text{Degrees of freedom} = 20.$$

Utilizing the relationships under equation (11) we obtain

$$\hat{\sigma} = 0.3907.$$

$$(2) \quad \ln(k_2) = 0.2336 + \underset{(0.2368)}{0.4608} \ln\left(\frac{\pi_L}{1-\pi_L}\right) ;$$

and with

$$R = 0.3909,$$

$$\text{Std. error est.} = 0.3865,$$

$$\text{Degrees of freedom} = 21.$$

so we get

$$\hat{\sigma} = 0.3154.$$

For the year 1890, the results are given by (3) and (4) as follows:

$$(3) \quad \ln(k_1) = 3.8687 + \underset{(0.1433)}{0.3849} \ln\left(\frac{\pi_L}{1-\pi_L}\right) ;$$

and with

$$R = 0.5055,$$

$$\text{Std. error est.} = 0.2604,$$

$$\text{Degrees of freedom} = 21.$$

so that

$$\hat{\sigma} = 0.2779.$$

$$(4) \quad \ln(k_2) = -0.2691 + \underset{(0.2671)}{1.1129} \ln\left(\frac{\pi_L}{1-\pi_L}\right) ;$$

and with

$$R = 0.6727,$$

$$\text{Std. error est.} = 0.4853,$$

$$\text{Degrees of freedom} = 21.$$

So we get

$$\hat{\sigma} = 0.5267.$$

For the year 1880, the results are given by (5) and (6) as follows:

$$(5) \quad \ln(k_1) = 3.7060 + \underset{(0.1489)}{0.1651} \ln\left(\frac{\pi_L}{1-\pi_L}\right) ;$$

and with

$$R = 0.2126,$$

$$\text{Std. error est.} = 0.4031,$$

$$\text{Degrees of freedom} = 26.$$

So we obtain

$$\hat{\sigma} = 0.1417.$$

$$(6) \quad \ln(k_2) = -0.2756 + \underset{(0.2686)}{0.2505} \ln\left(\frac{\pi_L}{1-\pi_L}\right) ;$$

and with

$$R = 0.1799,$$

$$\text{Std. error est.} = 0.7274,$$

$$\text{Degrees of freedom} = 26.$$

so that

$$\hat{\sigma} = 0.2003.$$

For the year 1870, the results are given by (7) and (8) as follows:

$$(7) \quad \ln(k_1) = 3.6477 + \underset{(0.1381)}{0.3185} \ln\left(\frac{\pi_L}{1-\pi_L}\right) ;$$

and with

$$R = 0.4190,$$

$$\text{Std. error est.} = 0.3665,$$

$$\text{Degrees of freedom} = 25.$$

so that

$$\hat{\sigma} = 0.2416.$$

$$(8) \quad \ln(k_2) = -0.3578 + \underset{(0.1810)}{0.5247} \ln\left(\frac{\pi_L}{1-\pi_L}\right) ;$$

and with

$$R = 0.5092,$$

$$\text{Std. error est.} = 0.4779,$$

$$\text{Degrees of freedom} = 24.$$

So we obtain

$$\hat{\sigma} = 0.3441.$$

For the year 1860, the result is given by (9) as follows:

$$(9) \quad \ln(k_2) = -0.2667 + \underset{(0.2247)}{0.2930} \ln\left(\frac{\pi_L}{1-\pi_L}\right) ;$$

and with

$$R = 0.2737,$$

$$\text{Std. error est.} = 0.4584,$$

$$\text{Degrees of freedom} = 21.$$

So we get

$$\hat{\sigma} = 0.2266.$$

B. The Wool Sector

For the year 1900, the result is given by (10) as follows:

$$(10) \quad \ln(k_1) = 3.1769 + \underset{(0.2924)}{0.1084} \ln\left(\frac{\pi_L}{1-\pi_L}\right) ;$$

and with

$$R = 0.0686,$$

$$\text{Std. error est.} = 0.5902,$$

$$\text{Degrees of freedom} = 29.$$

so that

$$\hat{\sigma} = 0.0978.$$

For the year 1880, the result is given by (11) as follows:

$$(11) \quad \ln(k_1) = 2.9049 + \underset{(0.1082)}{0.0649} \ln\left(\frac{\pi_L}{1-\pi_L}\right) ;$$

and with

$$R = 0.1071,$$

$$\text{Std. error est.} = 0.2723,$$

$$\text{Degrees of freedom} = 31.$$

So we get

$\hat{\sigma} = 0.0609.$

For the year 1870, the results are given by (12) and (13) as follows:

$$(12) \quad \ln(k_1) = \underset{}{2.9427} + \underset{(0.2131)}{0.3480} \ln\left(\frac{\pi_L}{1-\pi_L}\right);$$

and with

$R = 0.2814,$

Std. error est. $= 0.4956,$

Degrees of freedom $= 31.$

So we obtain

$\hat{\sigma} = 0.2582.$

$$(13) \quad \ln(k_2) = -0.7839 + \underset{(0.1441)}{0.8131} \ln\left(\frac{\pi_L}{1-\pi_L}\right);$$

and with

$R = 0.7117,$

Std. error est. $= 0.3353,$

Degrees of freedom $= 31.$

So that

$\hat{\sigma} = 0.4485.$

For the year 1860, the result is given by (14) as follows:

$$(14) \quad \ln(k_1) = 2.4212 + \underset{(0.1660)}{0.3075} \ln\left(\frac{\pi_L}{1-\pi_L}\right);$$

and with

$$R = 0.3474,$$

$$\text{Std. error est.} = 0.5143,$$

$$\text{Degrees of freedom} = 25.$$

It follows that

$$\hat{\sigma} = 0.2352.$$

We can see that all estimates for σ, for both sectors, are less than unity. This result coincides with our previous estimates. However, in most cases we can see that the regression coefficients are accompanied by significantly lower standard errors as compared with the previous estimates.

REFERENCES

Abbreviations

AER	American Economic Review
EHR	Economic History Review
EJ	Economic Journal
JEH	Journal of Economic History
JRSS	Journal of the Royal Statistical Society
OEEC	Organization for European Economic Cooperation
QJE	Quarterly Journal of Economics
UIB	University of Illinois Bulletin

Aldrich, Mr., Reports of the Committees of the Senate of the U.S. for the 2nd Session of the 52nd Congress, March 1893 (4 Vols.), Washington, D.C.: Government Printing Office, 1893.

Ames, E. and N. Rosenberg, "The Enfield Arsenal in Theory and History", EJ, Vol. 78 (December 1968) pp. 827-842.

Balfour, Sir Arthur, Chairman, Committee on Industry and Trade, Survey of Textiles Industries: Cotton, Wool, Artificial Silk, London: His Majesty's Stationery Office, 1928.

Blaug, M., "The Productivity of Capital in the Lancashire Cotton Industry During the Nineteenth Century", EHR, Series 2, Vol. 13 (1960-61) pp. 358-381.

Bowley, A.L., "The Statistics of Wages in the U.K. During the Last Hundred Years", JRSS, Vol. 65 (March 1902) pp. 102-126.

Bowley, A.L., Wages and Income in the U.K. Since 1860, London: Cambridge University Press, 1937.

Bowley, A.L., Wages in the United Kingdom in the Nineteenth Century, London: Cambridge University Press, 1900.

David, Paul A. and Th. Van Klundert, "Biased Efficiency Growth and Capital-Labor Substitution in the U.S. 1899-1960", AER, Vol. 55 (June 1965) pp. 357-394.

Davis, L.E. and J.R. Hughes, "A Dollar-Sterling Exchange 1803-1895", EHR, Series 2, Vol. 13 (1960-61) pp. 52-78.

Deane, P. and W. A. Cole, British Economic Growth 1688-1959: Trends and Structure, London: Cambridge University Press, 1962.

Drummond, Ian M., "Labor Scarcity and the Problem of American Industrial Efficiency in the 1850's: A Comment", JEH, Vol. 27 (September 1967) pp. 383-390.

Ellison, Thomas, The Cotton Trade of Great Britain (1st Edition), London: 1886.

Fellner, William, "Two Propositions in the Theory of Induced Innovation", EJ, Vol. 71 (June 1961) pp. 305-308.

Flux, A.W., "Industrial Productivity in Great Britain and the United States", QJE, Vol. 48 (November 1933) pp. 1-38.

Fogel, Robert W., "The Specification Problem in Economic History", JEH, Vol. 27 (September 1967) pp. 283-308.

Frankel, Marvin, British and American Manufacturing Productivity, UIB, Urbana: 1957.

Habakkuk, H. J., American and British Technology in the Nineteenth Century: The Search for Labor-Saving Inventions, London: Cambridge University Press, 1962.

Heath, J.B., "British-Canadian Industrial Productivity", EJ, Vol. 67 (December 1957) pp. 665-686.

Hicks, J.R., The Theory of Wages (1st Edition), London: McMillan, 1932.

Homer, Sidney, A History of Interest Rates, New Jersey: Rutgers University Press, 1963.

Layer, Robert G. and Arthur H. Cole, Earnings of Cotton Mill Operatives 1825-1914, Cambridge, Massachusetts: Harvard University Press, 1955.

McGouldrick, Paul F., New England Textiles in the Nineteenth Century, Cambridge, Massachusetts: Harvard University Press, 1968.

Mitchell, B.R. and P. Deane, Abstract of British Historical Statistics, London: Cambridge University Press, 1962.

Nerlove, Marc, "Recent Empirical Studies of the CES and Related Production Functions", in M. Brown (ed.) The Theory and Empirical Analysis of Production Studies in Income and Wealth, Vol. 31, National Bureau of Economic Research, New York, 1967.

Paige, Debora and Gottfried Bombach, *A Comparison of National Output and Productivity of the United Kingdom and the United States*, *OEEC*, Paris: 1959.

Rostas, L., *Comparative Productivity in British and American Industry*, London: Cambridge University Press, 1948.

Rothbarth, E., "Causes of the Superior Efficiency of U.S.A. Industry as Compared with British Industry", *EJ*, September 1946.

Samuelson, P.A., "Equalization by Trade of the Interest Rate Along with the Real Wage", in Essays in Honor of Gottfried Haberler, *Trade, Growth, and the Balance of Payments*, Chicago: Rand McNally, 1965.

Temin, P., "Labor Scarcity and the Problem of American Industrial Efficiency in the 1850's", *JEH*, September 1966.

Temin, P., "Labor Scarcity and the Problem of American Industrial Efficiency in the 1850's: A Reply", *JEH*, March 1968.

U.S. Bureau of the Census, *Historical Statistics of the United States, Colonial Times to 1957*, Washington, D.C.: Government Printing Office, 1960.

U.S. Bureau of the Census, *Historical Statistics of the U.S. 1789-1945: A Supplement to the Statistical Abstracts of the U.S.*, Washington, D.C.: Government Printing Office, 1949.

U.S. Bureau of the Census, *Statistical Abstract of the U.S. 1890*, Washington, D.C.: Government Printing Office, 1891.

U.S. Bureau of the Census, *Statistical Abstract of the U.S. 1895* Washington, D.C.: Government Printing Office, 1896.

U.S. Census Office, *Census Reports, Twelfth Census of the U.S. 1900, Manufactures: Textiles*, Washington, D.C.: Government Printing Office, 1902.

U.S. Census Office, *Census Reports, Vol. 7, Twelfth Census of the U.S. 1900, Manufactures Part 1*, Washington, D.C.: Government Printing Office, 1902.

U.S. Census Office, *Manufactures of the U.S. in 1860, Eighth Census*, Washington, D.C.: Government Printing Office, 1865.

Williamson, Harold F., "Mass Production, Mass Consumption and American Industrial Development", First International Conference of Economic History, Stockholm: 1960.

Dissertations in American Economic History

An Arno Press Collection

1977 Publications

Ankli, Robert Eugene. **Gross Farm Revenue in Pre-Civil War Illinois.** (Doctoral Dissertation, University of Illinois, 1969). 1977

Asher, Ephraim. **Relative Productivity, Factor-Intensity and Technology in the Manufacturing Sectors of the U.S. and the U.K. During the Nineteenth Century.** (Doctoral Dissertation, University of Rochester, 1969). 1977

Campbell, Carl. **Economic Growth, Capital Gains, and Income Distribution:** 1897-1956. (Doctoral Dissertation, University of California at Berkeley, 1964). 1977

Cederberg, Herbert R. **An Economic Analysis of English Settlement in North America, 1583-1635.** (Doctoral Dissertation, University of California at Berkeley, 1968). 1977

Dente, Leonard A. **Veblen's Theory of Social Change.** (Doctoral Dissertation, New York University, 1974). 1977

Dickey, George Edward. **Money, Prices and Growth;** The American Experience, 1869-1896. (Doctoral Dissertation, Northwestern University, 1968). 1977

Douty, Christopher Morris. **The Economics of Localized Disasters:** The 1906 San Francisco Catastrophe. (Doctoral Dissertation, Stanford University, 1969). 1977

Harper, Ann K. **The Location of the United States Steel Industry, 1879-1919.** (Doctoral Dissertation, Johns Hopkins University, 1976). 1977

Holt, Charles Frank. **The Role of State Government in the Nineteenth-Century American Economy, 1820-1902:** A Quantitative Study. (Doctoral Dissertation, Purdue University, 1970). 1977

Katz, Harold. **The Decline of Competition in the Automobile Industry, 1920-1940.** (Doctoral Dissertation, Columbia University, 1970). 1977

Lee, Susan Previant. **The Westward Movement of the Cotton Economy, 1840-1860:** Perceived Interests and Economic Realities. (Doctoral Dissertation, Columbia University, 1975). 1977

Legler, John Baxter. **Regional Distribution of Federal Receipts and Expenditures in the Nineteenth Century:** A Quantitative Study. (Doctoral Dissertation, Purdue University, 1967). 1977

Lightner, David L. **Labor on the Illinois Central Railroad, 1852-1900:** The Evolution of an Industrial Environment. (Doctoral Dissertation, Cornell University, 1969). 1977

MacMurray, Robert R. **Technological Change in the American Cotton Spinning Industry, 1790 to 1836.** (Doctoral Dissertation, University of Pennsylvania, 1970). 1977

Netschert, Bruce Carlton. **The Mineral Foreign Trade of the United States in the Twentieth Century:** A Study in Mineral Economics. (Doctoral Dissertation, Cornell University, 1949). 1977

Otenasek, Mildred. **Alexander Hamilton's Financial Policies.** (Doctoral Dissertation, Johns Hopkins University, 1939). 1977

Parks, Robert James. **European Origins of the Economic Ideas of Alexander Hamilton.** (M. A. Thesis, Michigan State University, 1963). 1977

Parsons, Burke Adrian. **British Trade Cycles and American Bank Credit:** Some Aspects of Economic Fluctuations in the United States, 1815-1840. (Doctoral Dissertation, University of Texas, 1958). 1977

Primack, Martin L. **Farm Formed Capital in American Agriculture, 1850-1910.** (Doctoral Dissertation, University of North Carolina, 1963). 1977

Pritchett, Bruce Michael. **A Study of Capital Mobilization, The Life Insurance Industry of the Nineteenth Century.** (Doctoral Dissertation, Purdue University, 1970). Revised Edition. 1977

Prosper, Peter A., Jr. **Concentration and the Rate of Change of Wages in the United States, 1950-1962.** (Doctoral Dissertation, Cornell University 1970). 1977

Schachter, Joseph. **Capital Value and Relative Wage Effects of Immigration into the United States, 1870-1930.** (Doctoral Dissertation, City University of New York, 1969). 1977

Schaefer, Donald Fred. **A Quantitative Description and Analysis of the Growth of the Pennsylvania Anthracite Coal Industry, 1820 to 1865.** (Doctoral Dissertation, University of North Carolina, 1967). 1977

Schmitz, Mark. **Economic Analysis of Antebellum Sugar Plantations in Louisiana.** (Doctoral Dissertation, University of North Carolina, 1974). 1977

Sharpless, John Burk, II. **City Growth in the United States, England and Wales, 1820-1861:** The Effects of Location, Size and Economic Structure on Inter-urban Variations in Demographic Growth. (Doctoral Dissertation, University of Michigan, 1975). 1977

Shields, Roger Elwood. **Economic Growth with Price Deflation, 1873-1896.** (Doctoral Dissertation, University of Virginia, 1969). 1977

Stettler, Henry Louis, III. **Growth and Fluctuations in the Ante-Bellum Textile Industry.** (Doctoral Dissertation, Purdue University, 1970). 1977

Sturm, James Lester. **Investing in the United States, 1798-1893:** Upper Wealth-Holders in a Market Economy. (Doctoral Dissertation, University of Wisconsin, 1969). 1977

Tenenbaum, Marcel. **(A Demographic Analysis of Interstate Labor Growth Rate Differentials;** United States, 1890-1900 to 1940-50. (Doctoral Dissertation, Columbia University, 1969). 1977

Thomas, Robert Paul. **An Analysis of the Pattern of Growth of the Automobile Industry:** 1895-1929. (Doctoral Dissertation, Northwestern University, 1965). 1977

Vickery, William Edward. **The Economics of the Negro Migration 1900-1960.** (Doctoral Dissertation, University of Chicago, 1969). 1977

Waters, Joseph Paul. **Technological Acceleration and the Great Depression.** (Doctoral Dissertation, Cornell University, 1971). 1977

Whartenby, Franklee Gilbert. **Land and Labor Productivity in United States Cotton Production, 1800-1840.** (Doctoral Dissertation, University of North Carolina, 1963). 1977

1975 Publications

Adams, Donald R., Jr. **Wage Rates in Philadelphia, 1790-1830.** (Doctoral Dissertation, University of Pennsylvania, 1967). 1975

Aldrich, Terry Mark. **Rates of Return on Investment in Technical Education in the Ante-Bellum American Economy.** (Doctoral Dissertation, The University of Texas at Austin, 1969). 1975

Anderson, Terry Lee. **The Economic Growth of Seventeenth Century New England:** A Measurement of Regional Income. (Doctoral Dissertation, University of Washington, 1972). 1975

Bean, Richard Nelson. **The British Trans-Atlantic Slave Trade, 1650-1775.** (Doctoral Dissertation, University of Washington, 1971). 1975

Brock, Leslie V. **The Currency of the American Colonies, 1700-1764:** A Study in Colonial Finance and Imperial Relations. (Doctoral Dissertation University of Michigan, 1941). 1975

Ellsworth, Lucius F. **Craft to National Industry in the Nineteenth Century:** A Case Study of the Transformation of the New York State Tanning Industry. (Doctoral Dissertation, University of Delaware, 1971). 1975

Fleisig, Heywood W. **Long Term Capital Flows and the Great Depression:** The Role of the United States, 1927-1933. (Doctoral Dissertation, Yale University, 1969). 1975

Foust, James D. **The Yeoman Farmer and Westward Expansion of U.S. Cotton Production.** (Doctoral Dissertation, University of North Carolina at Chapel Hill, 1968). 1975

Golden, James Reed. **Investment Behavior By United States Railroads, 1870-1914.** (Doctoral Thesis, Harvard University, 1971). 1975

Hill, Peter Jensen. **The Economic Impact of Immigration into the United States.** (Doctoral Dissertation, The University of Chicago, 1970). 1975

Klingaman, David C. **Colonial Virginia's Coastwise and Grain Trade.** (Doctoral Dissertation, University of Virginia, 1967). 1975

Lang, Edith Mae. **The Effects of Net Interregional Migration on Agricultural Income Growth:** The United States, 1850-1860. (Doctoral Thesis, The University of Rochester, 1971). 1975

Lindley, Lester G. **The Constitution Faces Technology:** The Relationship of the National Government to the Telegraph, 1866-1884. (Doctoral Thesis, Rice University, 1971). 1975

Lorant, John H[erman]. **The Role of Capital-Improving Innovations in American Manufacturing During the 1920's.** (Doctoral Thesis, Columbia University, 1966). 1975

Mishkin, David Joel. **The American Colonial Wine Industry:** An Economic Interpretation, Volumes I and II. (Doctoral Thesis, University of Illinois, 1966). 1975

Winkler, Donald R. **The Production of Human Capital:** A Study of Minority Achievement. (Doctoral Dissertation, University of California at Berkeley, 1972). 1977

Oates, Mary J. **The Role of the Cotton Textile Industry in the Economic Development of the American Southeast:** 1900-1940. (Doctoral Dissertation, Yale University, 1969). 1975

Passell, Peter. **Essays in the Economics of Nineteenth Century American Land Policy.** (Doctoral Dissertation, Yale University, 1970). 1975

Pope, Clayne L. **The Impact of the Ante-Bellum Tariff on Income Distribution.** (Doctoral Dissertation, The University of Chicago, 1972). 1975

Poulson, Barry Warren. **Value Added in Manufacturing, Mining, and Agriculture in the American Economy From 1809 To 1839.** (Doctoral Dissertation, The Ohio State University, 1965). 1975

Rockoff, Hugh. **The Free Banking Era: A Re-Examination.** (Doctoral Dissertation, The University of Chicago, 1972). 1975

Schumacher, Max George. **The Northern Farmer and His Markets During the Late Colonial Period.** (Doctoral Dissertation, University of California at Berkeley, 1948). 1975

Seagrave, Charles Edwin. **The Southern Negro Agricultural Worker:** 1850-1870. (Doctoral Dissertation, Stanford University, 1971). 1975

Solmon, Lewis C. **Capital Formation by Expenditures on Formal Education, 1880 and 1890.** (Doctoral Dissertation, The University of Chicago, 1968). 1975

Swan, Dale Evans. **The Structure and Profitability of the Antebellum Rice Industry:** 1859. (Doctoral Dissertation, University of North Carolina at Chapel Hill, 1972). 1975

Sylla, Richard Eugene. **The American Capital Market, 1846-1914:** A Study of the Effects of Public Policy on Economic Development. (Doctoral Thesis, Harvard University, 1968). 1975

Uselding, Paul John. **Studies in the Technological Development of the American Economy During the First Half of the Nineteenth Century.** (Doctoral Dissertation, Northwestern University, 1970). 1975

Walsh, William D[avid]. **The Diffusion of Technological Change in the Pennsylvania Pig Iron Industry, 1850-1870.** (Doctoral Dissertation, Yale University, 1967). 1975

Weiss, Thomas Joseph. **The Service Sector in the United States, 1839 Through 1899.** (Doctoral Thesis, University of North Carolina at Chapel Hill, 1967). 1975

Zevin, Robert Brooke. **The Growth of Manufacturing in Early Nineteenth Century New England.** 1975